New Frontiers in Regional Science: Asian Perspectives

Volume 74

This series is a constellation of works by scholars in the field of regional science and in related disciplines specifically focusing on dynamism in Asia.

Asia is the most dynamic part of the world. Japan, Korea, Taiwan, and Singapore experienced rapid and miracle economic growth in the 1970s. Malaysia, Indonesia, and Thailand followed in the 1980s. China, India, and Vietnam are now rising countries in Asia and are even leading the world economy. Due to their rapid economic development and growth, Asian countries continue to face a variety of urgent issues including regional and institutional unbalanced growth, environmental problems, poverty amidst prosperity, an ageing society, the collapse of the bubble economy, and deflation, among others.

Asian countries are diversified as they have their own cultural, historical, and geographical as well as political conditions. Due to this fact, scholars specializing in regional science as an inter- and multi-discipline have taken leading roles in providing mitigating policy proposals based on robust interdisciplinary analysis of multifaceted regional issues and subjects in Asia. This series not only will present unique research results from Asia that are unfamiliar in other parts of the world because of language barriers, but also will publish advanced research results from those regions that have focused on regional and urban issues in Asia from different perspectives.

The series aims to expand the frontiers of regional science through diffusion of intrinsically developed and advanced modern regional science methodologies in Asia and other areas of the world. Readers will be inspired to realize that regional and urban issues in the world are so vast that their established methodologies still have space for development and refinement, and to understand the importance of the interdisciplinary and multidisciplinary approach that is inherent in regional science for analyzing and resolving urgent regional and urban issues in Asia.

Topics under consideration in this series include the theory of social cost and benefit analysis and criteria of public investments, socio-economic vulnerability against disasters, food security and policy, agro-food systems in China, industrial clustering in Asia, comprehensive management of water environment and resources in a river basin, the international trade bloc and food security, migration and labor market in Asia, land policy and local property tax, Information and Communication Technology planning, consumer "shop-around" movements, and regeneration of downtowns, among others.

Researchers who are interested in publishing their books in this Series should obtain a proposal form from Yoshiro Higano (Editor in Chief, higano@jsrsai.jp) and return the completed form to him.

Yasuhiro Sakai

Games, Decisions, and Markets

Springer

Yasuhiro Sakai
Faculty of Economics
Shiga University
Hikone, Shiga, Japan

ISSN 2199-5974 ISSN 2199-5982 (electronic)
New Frontiers in Regional Science: Asian Perspectives
ISBN 978-981-99-5284-7 ISBN 978-981-99-5285-4 (eBook)
https://doi.org/10.1007/978-981-99-5285-4

This Springer imprint is published by the registered company Springer Nature Singapore Pte Ltd.
The registered company address is: 152 Beach Road, #21-01/04 Gateway East, Singapore 189721, Singapore

This book is dedicated to the memory of
Professor Oskar Morgenstern
Professor Asatoshi Maeshiro
Shiro, Toko, and Tora

Preface

Ze-a-mi (1363–1443), a legendary Japanese Noh drama writer-performer, once wrote the following maxim:

"Never forget the ideals with which you stated out."

Carefully looking back at my long career, it seems that I have never forgotten Ze-a-mi's maxim mentioned above. I was born in Osaka, one of the biggest cities in Japan, just before the Second World War broke out. Between 1943 and 1945, Osaka was bombed more than 30 times, and its damage and casualties caused by repeated bombing were very serious and widely extensive. Although a great number of buildings and private houses were destroyed by bombing, it was a miracle of miracles that my family barely escaped from the large-scale air raid. Then, when I was still a schoolboy, one of my respected teachers gave me lessons about the importance of holding fast to the lofty ideals of my youth. Since Japan was so poor and badly off, I determined to find a way to escape from misery, strongly wishing to become economics professor someday.

In my young student days at Kobe University, there existed two popular yet opposing textbooks in economics. They were nicknamed the "red text" and the "blue text." In hindsight, this was really a strange start for my student life. The red text was synonymous with the authoritative *Economics Textbook* by the Russian Academy Economics Institute. The issue of "socialism versus capitalism" constituted the central theme of the red text, declaring the proposition that "capitalism is destined to collapse whereas socialism is marching for its final victory over capitalism." In contrast to the powerful red text mentioned above, the blue text, represented by the popular book *Economics* written by the young American economist Paul A. Samuelson, seemed to be rather modest and even hesitant.

In those days, the red text seemed to give me more impact than the blue text. After some hesitation, however, I boldly decided to go over to United States to continue my graduate study without political and psychological interruptions. I just wanted to get out of the Japanese university disturbances in the 1960s, thus daring to jump into the very core of the capitalist economy.

Very fortunately, in 1968, I was admitted to the Graduate School of Economics, the University of Rochester. Professor Lionel W. McKenzie, a leading professor at Rochester, was one of those pioneers who together with Kenneth J. Arrow and Gerard Debreu greatly contributed to general equilibrium theory. In 1972, I finished my Ph.D. thesis in *Axiomatic Foundations of Consumption and Production Theories*, with McKenzie as my main advisor. Besides, thanks again to McKenzie's recommendation, I could get a teaching job at the University of Pittsburgh in 1971.

At Pittsburgh, I got acquainted with many helpful friends and hard-working students. Professor Asatoshi Maeshiro and his lovely wife Kazuko generously gave material and moral support to me and my wife Tokuko, i.e. the young couple starting a new life in America. After staying in Pittsburgh nicknamed Iron City, however, I gradually began to feel a considerable gap between Economic Theory and Economic Reality, finding increasing interest in Applied Economics. When Oskar Morgenstern, a famous pioneer of the theory of games, happened to give an inspiring lecture on the history of economic theories, he gave me kind advice for researching Games and Uncertainty.

In 1978, Toko and I safely returned to the Land of Rising Sun. We lived in the Tsukuba Academic Town, 100 km north of Tokyo. The striking contrast between New Citizens and Old Farmers gave us not only new and exciting experiences but also a series of unexpected frictions. We learned a lot from the old proverb *Adversity is the parent of virtue*. Then, we eventually settled down in Hikone, a small castle town near Kyoto, which is actually Toko's native place. Speaking of my life full of ups and downs for so many years, I have so frequently changed my place of residence in Japan and the United States, first from Osaka to Kobe, and next to Denver, Rochester, Pittsburgh, Hiroshima, Tsukuba, Tsuchiura, and finally to Hikone.

History sometimes does something cruel. A sequence of so many incredible things and events may happen beyond all expectations. In fact, the seemingly invincible Berlin Wall turned down all of a sudden in 1989, being followed by the total collapse of the mighty Soviet Union into so many independent countries in 1991. Until quite recently, I did not expect at all to return to Hikone and teach *Theory and History* at Shiga University and also at Ryukoku University, old private universities in Kyoto.

This time, when I was generously asked by Mr. Yutaka Hirachi and other staff members of Springer to write my third English book on economics, I racked my brains about what it should be all about. While the first book was entitled *J.M. Keynes Versus F.H. Knight: Risk, Probability, and Uncertainty* and published in 2019, the second book coauthored with my former student Dr. Keisuke Sasaki, Toyo University, was entitled *Information and Distribution: The Role of Merchants in the Market Economy under Uncertainty*, and safely published in 2021, just 2 years ago. In my opinion, the third book should be somehow related to, yet completely independent of, the first and second books. Upon mature consideration, I have determined that the title of the third new book should be *Games, Decisions, and Markets*, which reflects kind and consistent advices from so many scholars including Oskar Morgenstern, Asatoshi Maeshiro, Lionel W. McKenzie, James W. Friedman,

Paul A. Samuelson, Hirofumi Uzawa, Masayoshi Hirota, Edmand Malinvaud, John Hicks, Michio Morishima, Ronald Jones, and Murray Kemp.

This book consists of three parts. Broadly speaking, Part I deals with Games, containing the first three chapters, namely Chaps. 1, 2, and 3. Part II turns to Decisions, consisting of the following four chapters, that is, Chaps. 4, 5, 6, and 7. And Part III is concerned with Markets, being composed of the final two chapters, i.e., Chaps. 8 and 9. In what follows, more detailed discussions of each chapter will be given.

Part I is expected to have a threefold purpose. First of all, it aims to serve as a comprehensive introduction to the whole structure of the Theory of Games, which was first established by the joint work of the famous mathematician John von Neumann and the noted economist Oskar Morgenstern. Second, it systematically discusses the Theory of Games with the help of many graphical illustrations. Third, it specifically deals with zero-sum and non-zero-sum games, with special reference to literary works of Conan Doyle, Edgar Alan Poe, and many others.

To begin with, Chap. 1 gives a historical approach to the Theory of Games. There are two memorial years for the theory—1928 as the Year of Birth and 1944 as the Year of Maturity. It is an extremely important matter to distinguish accurately between "a game under risk" discussed in Conan Doyle's detective story *The Final Problem* and "a game under uncertainty" shown by Doyle's story *The Adventure of the Empty House*. I believe that the second kind of games, in which "animal spirits" à la John M. Keynes play a key role in decision-making, remain to be fairly unexplored, requiring future research.

Chapter 2 continues to critically reassess the significance and limitations of zero-sum games. First, I intensively discuss several zero-sum two-person games, with special reference to the Games of Matching Pennies. New graphical illustrations are attempted for clarification of the matter. Then, I turn to novelist Edgar Alan Poe's best story *The Purloined Letter*, showing how the Poe story may challenge the validity of Neumann-Morgenstern Approach. Next, I shed new light on Conan Doyle's story *The Final Case* as a variant of the Games of Matching Pennies. Finally, I pick up the Game of Stone, Paper, Scissors, which constitutes another interesting example of zero-sum two-person games.

Chapter 3 critically discusses non-zero-sum games with special reference to Nash equilibrium, which can be regarded as a successor of Cournot equilibrium in the theory of duopoly and oligopoly. I first focus on the "Residence Game," whose fictional counterpart has often been written in contemporary novels. I pick up the old and the young couples, who have to decide whether they live together or separately, and whether they live in the country or in the city. Now I shed light on "Generation Gaps Problems," with many other applications. I next turn to the "Battle of the Sexes," or plainly the "Dilemma of Lovers." The male and the female have to find good dating spots. While the male prefers to see boxing rather than ballet, the female's preference is just the opposite. I propose that contrary to the conventional way of reasoning, the introduction of a *third opinion* such as "going to see movies" is good enough to give each couple a *second-best solution*. Finally, comparison between "Econs" and "Humans" is carefully examined.

Part II carefully explores the foundations of consumer and producer decision theories, which constitute the "two keystones" in the grand structure of micro-economic theory. Although its earliest version was my Ph.D. thesis a long time ago, its completely revised draft toward new policy orientation is presented here. Besides, many instructive figures are newly drawn for better understanding.

Chapters 4 and 5 argue the foundations of consumer decision. First, Chap. 4 discusses consumer decision and revealed preference from many angles, thus carefully reexamining the foundations of economic analysis in modern times. It is Paul A. Samuelson's contribution who boldly introduced to micro-economic theory the brand-new concept of *revealed preference* against the then current doctrine of ordinal utility. The main result of this chapter is that Houthaker's *strong axiom of revealed preference* holds if and only if Samuelson's *weak axiom of revealed preference* and Sakai's *regularity condition* both hold. This new equivalence result distinguishes itself from all the previous works in that neither continuity nor Lipschitz assumptions are made on the demand function. Finally, in the light of uncertainty and behavioral economics in the present times, the significance and limitations of Samuelson-type economics are also discussed.

Chapter 5 studies the foundations of the *indirect utility function*, based on a revealed preference approach à la Paul A. Samuelson. I look at a chain of comparisons of budgets as if it gives a *relation on the normalized-price space* (namely, a *revealed favorability relation*) rather than a relation on the commodity space (namely, a revealed preference relation). In analogy to the weak and strong axioms of revealed preference, the weak and strong axioms of revealed favorability are newly introduced, and a fundamental theorem concerning the relationship between the latter two axioms is skillfully established. Then, the indirect and direct utility functions are effectively derived on the basis of the strong axiom of revealed favorability. It is noted that *neither the continuity of the demand function nor the convexity of its range is required for the approach taken here.*

Chapters 6 and 7 discuss the foundations of producer decision. The purpose of Chap. 6 is to shed new light on input demand theory, which has unfortunately been paid less attention than consumer demand theory. I start our discussion with general production possibility sets rather than specific production functions, driving what we may call *decomposition equations in input demand theory*. The total effect of a change in the price of a certain input on the demand for another input can be divided into the following two separate effects. They are a *substitution effect along the old isoquant* and *an expansion effect along the new expansion path.* Such a decomposition equation in input demand theory appears to correspond well to the famous Slutsky equation in consumer demand theory. The correspondence between the producer's decomposition and the consumer's decomposition, however, is not perfect. How and to what extent they are different should be actually the target of my investigation. In particular, I pay special attention to comparison between the producer's expansion effect and the consumer's income effect. In this connection, the effectiveness of LeChatelier-Samuelson principle is also confirmed.

Chapter 7 aims to extend our new approach to input demand theory with a single output to the more general case of *joint production.* Although such an extension from one output to several outputs seems to be straightforward, it nevertheless requires special care and careful interpretation. More specifically, we are concerned with a systematic examination of the question what happens when the producer that has been in equilibrium at certain prices of inputs and prices of outputs experiences a change in one of those prices. *The Mutual Effects between inputs and between outputs and the Cross Effects between inputs and outputs* are newly explored by means of *various kinds of decomposition equations in production theory.* A definition of the *normal technology* is given to show that inputs are not gross substitutes, nor are outputs, and that the input–output relations are not regressive. This is actually an extension of the Trout Rader result to the case of joint production.

Part III shifts again my investigation from micro-economics to general equilibrium theory. More specifically, close interdependence of several markets at one country level, and international trade model in the presence of risk are explored both intensively and extensively.

Chapter 8 aims to shed new light on the Hicks-Morishima approach to the interdependence of several markets. In spite of its rather simple and ambitious framework, it is quite unfortunate that this approach has been rather neglected in the academic circle. I think that there must exist several reasons for such underestimation. First, the standard general equilibrium approach developed by Lionel W. McKenzie, Gerald Debreu, and Kenneth W. Arrow exclusively works with the commodity space rather than the price space. In contrast, the Hicks-Morishima approach based on Hicks' classical book *Value and Capital* uniquely *operates on the price space*, thus against the current mainstream of economic theory. Next, the majority of economics readers are usually familiar with straightforward notion of demand and supply curves, but not with the *twisted concept of excess demand curves.* It is one of my main purposes to mend such unlucky tendency, thus proceeding toward the establishment of a *new grand system of social science.*

The risk-free, two-sector, constant-returns-to-scale model has long served as a standard model of international trade, with Ronald W. Jones and Murray C. Kemp being its key promoters. The purpose of Chap. 9 is to make an effort to extend the basic theorems of the standard model to *cover new situations with price risk.* The question to ask is whether and to what extent those results are still applicable to the *new world under risk.* It is shown that when firms in the stochatic sector exhibit decreasing absolute risk aversion, the Rybczynski theorem and the Stolper-Samuelson theorem may fail to hold for some important cases, whereas the factor-price equalization theorem cannot be carried over to the stochastic world at all. The basic reason for such annoying results is actually the presence of *risk-bearing fee* associated with price risk. In short, when any form of risk or uncertainty is introduced into our model, many of the traditional results may lose their validity. In the real world, uncertainty really matters.

The preparation of this book has been made possible by the Japanese Ministry of Education, Culture, Sports, Science and Technology through Grand-in-Aid for Scientific Research. Once again, I would like to say my special thanks to Professor Yoshiro Higano (Editor-in-Chief), Professor Hirotada Kono (Honorary Editor), Professor Makoto Tawada (Managing Editor), Professor Peter Nijkamp (Chair of

Advisory Board), and many other Editorial and Advisory Board Members for generous support and nice suggestions. Last but not least, I am also very grateful to Mr. Yutaka Hirachi (senior editor), Ms. Kavitha Jayakumar, Ms. Pavitra Arulmurugan, Ms. Misao Taguchi, and other staff members of Springer Nature for generous assistance and heartwarming encouragement. Thank you so much.

Hikone, Shiga, Japan
1 June 2023

Yasuhiro Sakai

Contents

Part I
Games

Chapter 1
Von Neumann, Morgenstern, and Theory of Games: Critical Reassessment of Zero-Sum Games

Abstract This chapter gives a brief yet critical account of the theory of games as jointly developed by the two superstars in the twentieth century; namely, gifted mathematician John von Neumann and brilliant economist Oskar Morgenstern. There are two memorial years for game theory: 1928 as the Year of Birth for Game Theory and 1944 as the Year of Maturity for Game Theory. Even after the basic mathematical skeleton for game theory was provided by von Neumann in his 1928 paper, giving its real body and substance to the skeleton was a difficult job for Morgenstern. In both his 1928 paper and 1935 paper, Morgenstern paid a special attention on the duel between Sherlock Holmes and Professor Moriarty in Conan Doyle's famous detective story, finding an unsolved puzzle between an infinite chain of "out-thinking" and a finite concept of general equilibrium. In 1944, the collaboration of von Neumann and Morgenstern finally produced a difficult bulky book *Theory of Games and Economic Behavior*, whose academic reaction was initially rather quiet, requiring many years for its full recognition by social and natural scientists. It is of utmost importance to draw a definite line between "games under risk," represented by *The Final Problem* in Doyle's detective stories, and "games under uncertainty," shown by *The Adventure of the Empty House*. We believe that the second kind of games, in which "animal spirits" a la John M. Keynes play a key role in decision-making, remain to be fairly unexplored, requiring for future research.

Keywords John von Neumann · Oskar Morgenstern · Game theory · The duel between Sherlock Holmes and Professor Moriarty · Sherlock Holmes versus Colonel Moran · Games under risk · Games under uncertainty · The role of animal spirits

This chapter is an enlarged English version of Sakai (2001), with new figures. I am grateful to Oskar Morgenstern (Princeton), James Friedman (Rochester), Edward Zabel (Rochester), and Mitsuo Suzuki (Tokyo Kogyo) for helpful suggestions.

Y. Sakai, *Games, Decisions, and Markets*, New Frontiers in Regional Science: Asian Perspectives 74, https://doi.org/10.1007/978-981-99-5285-4_1

1.1 "Specter of Game Theory": An Introduction

The Manifesto of the Communist Party (German original: *Manifest der Kommunistischen Partei*) was an 1848 pamphlet published as the joint work of two giants in philosophy and social science: Karl Marx (1818–1883) and Friedrich Engels (1820–1895). Although it was a rather small pamphlet containing only 100 pages or so, it has been very influential in the economic profession until today. Remarkably, it had a very famous introduction beginning with the following sentence:

> A specter is haunting in Europe—the specter of communism. (Marx and Engels 1848, p. 1)

It seems that a similar phenomenon is taking place in modern economics today. On the analogy of *the Communist Manifesto*, we can metaphorically express such a new phenomenon as follows:

> A specter is haunting in the Economics profession the specter of Game Theory.

Interestingly enough, the theory of games, or more simply game theory, was also the joint product of two superstars in science: John von Neumann (1903–1957) and Oskar Morgenstern (1902–1977). The mathematician Neumann and the economist Morgenstern successfully combined their knowledge and skill to a revolutionary theory of games and economic behavior, based on the interactions of strategies of many players. When the final product of their collaboration appeared as a bulky book *Theory of Games and Economic Behavior* (von Neumann and Morgenstern 1944), namely the year near to the end of the long and dreadful Second World War, an authoritative mathematical journal wrote the following impressive sentence in an ecstasy of joy:

> Posterity may regard this book as one of the major scientific achievements of the first half of the twentieth century. (*American Mathematical Society Bulletin* 1944)

It seems, however, that Mitsuo Suzuki (1999) had an entirely different opinion of the new theory of game. Although he himself was one of the outstanding pioneers of game theory in Japan, he once lamented over the effectiveness and applicability as follows:

> When I [Suzuki] started doing research in game theory, my work was exposed to criticism from Japanese academia. To be honest, I undergo a neglect, or even attack, from so many people for my "apparently fruitless effort." Then I had to practice my perseverance, sincerely hoping that my lonely effort would be rewarded someday in the not too distance future. (Suzuki 1999, p. 7)

In retrospect, the theory of games was once ignored by so many people and even by feared by some people. There were a number of reasons for such unjustifiable treatment in the academia. First of all, the name "game theory" per se appeared to be provocative and amusing, thus spreading the wrong impression among the general public. In everyday conversation, if we accuse some persons of "playing games," we mean that they are not serious enough about a difficult situation, or that they are deliberately misleading us or making us do unnecessary things. For example, people can use trump cards to enjoy many games such as poker and bridge.[1]

It is quite remarkable to see that in commemorating the fifth anniversary of the publication of *Theory of Games and Economic Behavior* (1944), the Princeton University Press made the following announcement:

> A great book often requires so many years for its general recognition. When it is finally recognized as such by the public, its influential range will far-exceed the mere readership, thus appealing to the whole society. (*American Scientist* 1949, quoted by Poundstone 1992, p. 63)

To be honest, only 4000 volumes were sold for 5 years from 1944. The original Neumann–Morgenstern book was not only scarcely read by professional economists, but also almost ignored by many libraries. Interestingly enough, however, it was enthusiastically bought by some professional gamblers.

Second, the contents of the book were filled with the apparently strange combination of powerful mathematics and special economics, with strange equations and odd figures/tables. As a result, so many researchers who had a habit of sweeping a troublesome problem under the carpet displayed strong risk aversion to game theory.

Furthermore, the book itself was written in a sort of "German English," definitely not in standard British English. Since von Neumann was born in Budapest, the Kingdom of Hungary, and Oskar Morgenstern in a small town of the German Empire, their communications were usually done in German rather than English.[2]

Because of those reasons aforementioned, game theory was born as "an unfortunate child." We must add, however, that when grown up, such "a luckless child" became "a great figure of many talents." The purpose of this chapter is to briefly discuss how such a great transformation has been accomplished for those long years. As the saying goes, Rome was not built in a day.

The contents of this paper are as follows. In the second section, we will discuss a short history of theory of games, from its lonely birth to popular maturity. It is really important as well as interesting to know how the collaboration between von Neumann and Morgenstern, two distinguished scholars in different fields and from different countries, became possible in the times of hardships and wars. The third

---

[1] For the sad history of game theory in its early days, Mitsuo Suzuki once lamented it as follows: "The theory of games was born and regarded long as heretical. Its wide recognition as an important contribution to science took really long years" (Suzuki 1999, p. 7). Suzuki has been a lone pioneer in game theory in Japan and worked hard as a professor at Tohoku University and Tokyo Institute of Technology. Fortunately, his long and lonely struggle is now amply rewarded.

[2] The original book *Theory of Games and Economic Behavior* (1944) was a very bulky book, full of mathematical equations and strange notations. It was treated with "fear and respect." In commemorating the fifth anniversary of its publication, the Princeton University Press voiced the following announcement in *The American Scientist*: "Recognition of first-rate books often requires so many years. Once it is rightly recognized as such, however, its influence will far-exceed the narrowed scope of readers, probably reaching the general public in the world." To tell the truth, the volume of the first edition was only 4000. Understandably, it was hardly read by economists and purchased by a limited number of libraries. Remarkably, however, it was bought by some curious gamblers. For details, see Poundstone 1992, p. 63.

section will repeatedly confirm that Arthur Conan Doyle's novel featuring Sherlock Holmes as a famous detective constantly plays a key role for the birth and development of the revolutionary theory of strategies and games. It will be shown that the famous duel between Sherlock Holmes and Professor Moriarty gives us the original form of standard "zero-sum two-person games." As the saying goes, it is really a "mission impossible" to create something from nothing. Several remarks will be made in the final fourth section.

1.2 The Two Memorable Years for Game Theory: 1928 as the "Year of Birth" and 1944 as the "Year of Maturity"

This section will outline the birth and development of game theory in a historical perspective. As is seen in Table 1.1, there are the two memorable years for game theory: first, 1928 as the "Year of Birth", and second, 1944 as the "Year of Maturity."

Table 1.1 John von Neumann and Oskar Morgenstern: their lives and collaboration on game theory

<table>
<tr><th>John von Neumann</th><th>Oskar Morgenstern</th></tr>
<tr><td>1903 Born in Budapest, Hungary. His father was a wealthy Jewish banker</td><td>1902 Born in Görlitz, Germany. His mother was said to be a noblewoman</td></tr>
<tr><td>1928 "Theory of social games" (the German original), this was historically the first mathematical paper on game theory</td><td>1928 "Economic forecasting" (the German original). The Holmes-Moriarty duel was first noticed as a "troublesome game"</td></tr>
<tr><td colspan="2">≫1928 was regarded as the YEAR OF BIRTH for game theory</td></tr>
<tr><td>1933 Appointed as a lifetime professor at the Institute for Advanced Study, Princeton University, New Jersey</td><td>1935 "Perfect foresight" (the German original). The Holmes-Moriarty duel was again noted as an obstacle to perfect foresight. Then, Edward Chech pointed Morgenstern to Neumann's paper (1928) above</td></tr>
<tr><td>1937 Became a naturalized citizen of the U.S.</td><td>1938 While Morgenstern was visiting American universities, the Nazis took over in Vienna</td></tr>
<tr><td colspan="2">(Both Neumann and Morgenstern stayed at Princeton)</td></tr>
<tr><td colspan="2">1939 After giving an after-luncheon talk, Morgenstern talked with Neumann about game theory. Their talk continued for a long time</td></tr>
<tr><td colspan="2">1940 Morgenstern wrote a greatly enlarged version of Neumann's old paper (1928). When Neumann saw it, he suggested to Morgenstern: "Why don't we write this paper together?" Their earnest collaboration began</td></tr>
<tr><td colspan="2">1944 The first edition of Theory of Games and Economic Behavior was published by Princeton University Press
≫Thus, regarded as the YEAR OF MATURITY for game theory</td></tr>
<tr><td>1957 Neumann passed away</td><td>1977 Morgenstern passed away</td></tr>
</table>

To begin with, we will focus on the first memorable year of 1928. Historically speaking, this is the year in which the world economy reached the height of its prosperity. Although many European countries suffered a great deal from the aftermath of the First World War, they gradually recovered in the late 1920s, so that many people really began to dream of the everlasting capitalist economy. As the old ballade often tells us, however, all things are uncertain and must pass. In 1928, a year earlier than the outbreak of the Great Depression, ordinary people enjoyed their daily lives, being apparently confident of the continuation of their "golden days."

It is in this 1928 that von Neumann, a native son of Budapest, the Kingdom of Hungary, wrote the following outstanding paper on "social game theory";

> von Neumann (1928), "Zur Theorie der Gesellschaftsspiele," (English translation: The theory of social game), *Mathematische Annalen*, Vol. 100.

There are two things noticeable. First, this was an old technical paper that was written in 1928, with the language used being German, but not English. Second, although it discussed people's social games, it was published in a mathematical journal, thus representing the unique combination of social science and mathematics.[3]

John von Neumann was born in Budapest, Kingdom of Hungary, which was then an integral part of the Austro-Hungarian Empire. He was the eldest son of a wealthy and nonobservant Jewish family, with his father being an influential banker. He himself was a very clever boy with bright mathematical skill and often called "child prodigy." While by the early age of 8, he was familiar with differential and integral calculus, he found special interest in history as well. He started his lectures as a *Privatdozent* at the University of Berlin in 1928. On New Year's Day in 1930, he married Marietta Kövesi at Budapest University. Von Neumann and Marietta had one child, a daughter, Marina, born in 1935. In the early 1970s, Yasuhiro Sakai became an assistant professor of mathematical economics at the University of Pittsburgh. Very fortunately, Sakai soon got acquainted very well with Maria, who was then highly respected as Professor Marina von Neumann Whitman, eagerly teaching international economics. In hindsight, it was really amazing to trace the existence of "Econ Connection" in mathematician von Neumann's early career through both his first wife and his only daughter.[4]

---

[3]For details, see Aumann and Hart (1992), Dasgupta et al. (1992), Dixit and Nalebuff (1991), Luce and Raiffa (1957, rev. 1989), Nakayama (1997), Nakayama et al. (2000), Okada (1996, rev. 2011), Sakai (2001), Shubik (1967, 1982), and Varoufakis (2001).

[4]For the life and work of John von Neumann, see Poundstone (1992). Neumann has only one daughter, whose name is Marina von Neumann Whitman. When Marina was very young, she was a very clever girl, and said to George Gamow, a well-known science writer, that other than pure mathematics, she was much clever than her father, John von Neumann. And she added that even in pure mathematics, she was almost as clever as John (see Preface, Gamow 1947, rev. 1961). In the early 1970s, Sakai was an assistant professor of mathematical economics at the University of Pittsburgh. Sakai still has a very fond memory of Marina, already a noted professor of international economics, who once told Sakai that she had lost much interest in pure and applied mathematics.

In 1933, von Neumann was offered a lifetime professorship at the Institute for Advanced Study, Princeton University. In Princeton, he liked to play loud German march music on his phonograph, annoying his neighbors including Albert Einstein, the creator of the theory of relativity. It seemed that even in the United States, Neumann was fond of German culture and music except "Nazi connection," enjoying lively conversation with his friends in German.

To put it perhaps a little strongly, Oskar Morgenstern, the hard-working economist whose destiny was to become von Neumann's good collaborator, seemed to be influenced by German culture more strongly than von Neumann. Morgenstern was born in Görlitz, Germany, a city near the border with Poland. He grew up in Vienna, Austria, and graduated from the University of Vienna and got his Ph.D. in political science, later becoming a professor in economics at the same university in 1928. During his visit to Princeton University in 1938, something he did not expect happened all of a sudden: Adolf Hitler took over Vienna and expanded his political and military influence over the whole Europe. Since Morgenstern was unfairly dismissed as "politically unbearable" from the University of Vienna, he decided to remain in the United States. He became a member of the faculty at Princeton University and particularly gravitated toward the Institute for Advance Study in which von Neumann already stayed as a lifetime professor.[5]

In February 1939, Morgenstern gave an after-luncheon talk on the business cycles at the Nassau Club, von Neumann happened to be there with Niels Bohr, a distinguished particle physicist from Denmark. Both von Neumann and Bohr invited Morgenstern that afternoon for tea at Fine Hall, and all of them spent several enjoyable hours talking about games and experiments. Afterward, Morgenstern and von Neumann had many opportunities to talk about wide-ranging topics. One day, Morgenstern mentioned to von Neumann that Morgenstern was very much interested in studying von Neumann's (1928) paper on game theory. When Morgenstern had a chance to talk to von Neumann about his pet work on incompatibility between perfect foresight and social games, he had a rather unexpected reply from von Neumann. In all honesty, von Neumann told Morgenstern that von Neumann had done no further work on game theory for those many years after 1928.[6]

This made Morgenstern to study von Neumann's paper on game theory very seriously. More specifically, Morgenstern referred to the episode of the pursuit of Sherlock Holmes by Professor Moriarty in Conan Doyle's famous detective story. Morgenstern explained to von Neumann in some detail that Holmes and Moriarty could never be resolved on the basis of one of them "out-thinking" the other. The

[5]For the life of Oskar Morgenstern and his collaboration with von Neumann, see Morgenstern (1976) and Suzuki (1994). Also see Morgenstern (1958, 1972).

[6]By chance, the year of 1928 was the Year of the Dragon in terms of the Japanese animal calendar. In that memorial year, we saw the birth of many famous Japanese economists including Hirofumi Uzawa as a worldwide pioneer of general equilibrium theory, growth theory, and the economics of global warming, and Mitsuo Suzuki as an outstanding promoter of game theory and its applications. Twelve years later than Uzawa and Suzuki, Sakai was born in 1940, another Year of the Dragon.

infinite chains of "out-thinking," well illustrated by the never-ending sentence "if Mr. Holmes thinks that Mr. Moriarty thinks that Mr. Holmes thinks. . .," would guide the reader to a blind alley with no hope to escape. He showed that the strong assumption of "perfect foresight" would lead to inescapable paradoxes, thus being inadmissible for a seemingly powerful general equilibrium theory.[7]

Morgenstern had then started to investigate many possibilities and further developments of game theory. Morgenstern expanded game theory in a variety of ways, so that their joint research papers were greatly accumulated. Then, von Neumann himself suggested to Morgenstern as follows: "Why don't we write this paper together?" Morgenstern solemnly accepted this proposal right away. Initially, their joint work on game theory was prearranged as a mere pamphlet of around 100 pages. As their joint work progressed with the years, however, the number of pages of the pamphlet unexpectedly inflated.

September 18, 1944, was really a very historical moment for game theory! At this particular moment when the long and destructive Second World War was closing at the end, the product of the long and exciting collaboration of von Neumann and Morgenstern was finally released as a quite bulky book with more than 600 pages. So, the year of 1944 should rightly be celebrated as the Year of Maturity for Game Theory.

1.3 How to Solve "The Final Problem": Conan Doyle's Solution and Oskar Morgenstern's Objection

If we take a close look at the birth and development of the theory of games, then we can immediately learn that one famous detective story has made a very outstanding contribution to it. That is no less than the short story *The Final Problem* by Conan Doyle, featuring his detective character Sherlock Holmes. It was first published in *The Strand Magazine* in the United Kingdom, under the title *The Adventure of the Final Problem* in December 1893. Then, it reappeared in book form as a part of the collection *The Memoirs of Sherlock Holmes* (Doyle 1894). For details, see Doyle (1894, 1903).

Historically speaking, writer Conan Doyle (1859–1930) was more than 40 years older than mathematician John von Neumann (1903–1957) and economist Oskar Morgenstern (1902–1977). Although famous detective Sherlock Holmes was a fictitious character created by Doyle himself, it had long been a superstar in the

[7] There were so many economists who got quite impressed by Morgenstern's German paper (1935). Among those people were Frank H. Knight, a pioneer of the economics of uncertainty at Chicago, who took pains to translate it to English and fondly use it as a material in his lecture. Its official publication, however, was done far later in Schotter (ed.) (1976) *Selected Economic Writings of Oskar Morgenstern*. Its English paper was titled "Perfect Foresight and Economic Equilibrium." As the saying goes, life is short but art is long! A further investigation between Knight and Morgenstern should be eagerly awaited.

general public, well known over the British Channel throughout the European Continent. It was Doyle's original plan that the story *The Final Problem* was literally "Holmes's Final Problem," meaning the death of leading character Sherlock Holmes. In reality, however, the schedule is no more than a schedule and often apt to be broken.[8]

As soon as the detective story *The Final Problem* was published together with the implication of the sudden death of Sherlock Holmes, a very strong storm of protest from the readers directed against the author Conan Doyle: "Our beloved detective Holmes should be an invincible man. Mr. Doyle, please let Holmes come back!" Doyle had constantly been under strong pressure for the revival of Holmes by any possible means. So, after 10 long years since the publication of *The Final Problem* in 1893, Doyle wrote the new adventure story *The Adventure of the Empty House* in *The Strand Journal*, the October issue, 1903. This was the reason why the 1905 third collection of short stories was entitled *The Return of Sherlock Holmes*.

Oskar Morgenstern was then fond of reading Doyle's detective stories, thus becoming a worldwide member of "Sherlockian Group" or the group of Sherlock Holmes fans. Believe it or not, Morgenstern already introduced Doyle's detective story *The Final Problem* two times before it was again appeared in *Theory of Games and Economic Behavior* (1944), a monumental work with von Neumann.

Let us discuss this point in detail. As was seen in Table 1.1, Morgenstern (1928) first discussed *The Final Problem* in his German book *Wirtschaftsprognose* (English: *Economic Forecasting*). Interestingly enough, Morgenstern (1935) again referred to the same problem in his article *Vollkommene Voraussicht und Wirtschaftliches Gleichgewicht* (English translation: *Perfect Foresight and Economic Equilibrium*) in a prestigious German journal. And astonishingly, in von Neumann and Morgenstern (1944), *The Final Problem* was once again taken up for careful discussion, playing a critical role in the development of game theory. Many movie fans may well remember the following title: *007 Is Alive Twice*. By using a similar expression, we could say the following: *The Final Problem* seems to be alive three times! History repeats itself. Likewise, a famous story may repeat itself!

As the saying goes, seeing is believing. *The Final Problem* was concerned with the life-or-death struggle between famous detective Sherlock Homes and mathematical evil Professor Moriarty. Holmes described Moriarty as follows:

> He [Moriarty] is the Napoleon of crime, Watson. He is the organizer of half that is evil and of nearly all that is undetected in this great city. He is a genius, a philosopher, an abstract thinker. He has a brain of the order. (Doyle 1894, pp. 335–336)

[8]Writer Conan Doyle was not born in England, but in Scotland, an archrival of England. Interestingly enough, he graduated from the Faculty of Medicine, University of Edinburgh. By chance, Adam Smith, called the Father of Economics, was also a son of Scotland, being a graduate of the University of Glasgow. Scotland used to be the home of Celtic culture, which seems to do some contribution to the Study of Man.

The struggle between Sherlock Holmes and Professor Moriarty was first introduced in Morgenstern (1928) and reproduced in Morgenstern (1935). It was actually Morgenstern's favorite theme long before he met von Neumann and even after their collaboration started at Princeton. First of all, we would like to note the following passage from Morgenstern (1928), which was later reproduced in Morgenstern (1935):

> Sherlock Holmes, pursued by his opponent, Moriarty, leaves London for Dover. The train stops at a station on the way, and he alights there rather than traveling on to Dover. He has seen Moriarty at the railway station, recognizes that he is very clever and expects that Moriarty will take a faster special train in order to catch him in Dover. Holmes' anticipation turns out to be correct. But what if Moriarty had been still more clever, had estimated Holmes' mental abilities better and had foreseen his action accordingly? Then, obviously, he would have traveled to the intermediate station. Holmes, again, would have had to calculate that, and he himself would have decided to go on to Dover. Whereupon, Moriarty would again have "reacted" differently. Because of so much thinking they might not have been able to act at all or the intellectual weaker of the two would have surrendered to the other in the Victoria Station, since the whole flight would have become unnecessary. (Morgenstern 1928, Morgenstern 1935; quoted in Schotter 1976, pp. 175–176)

When Morgenstern wrote this passage, he served as Professor of Economics at the University of Vienna. As a professional economist, he knew very well what "economic equilibrium" was all about. On the one hand, in the duel between Holmes and Moriarty, each person has to "out-think his opponent" in the form "I think that he thinks that I think..." This is a sort of never-ending sequence of predictions. On the other hand, on ordinary economic equilibrium, the demand and supply of any good are adjusted at the market. Once the market equilibrium is realized, the quantities of trading are determined once and for all, with no need of "out-thinking" and "manipulation." Consequently, it was no wonder that Morgenstern found a sort of "insolvable paradox" between "the Holmes-versus-Moriarty way of out-thinking" and "the traditional way of trading in the market." In this connection, let us quote the following passage from Morgenstern (1935):

> One may be easily convinced that here lies an insoluble *paradox*. And the situation is not improved, but, rather, greatly aggravated if we assume that more than two individuals as, for example, in the case with exchange are brought together into a position, which would correspond to the one brought forward here; Always, *there is exhibited a endless chain of reciprocally conjectured reactions and counter-reactions. This chain can never be broken by an act of knowledge but always only through an arbitrary act a resolution.* This resolution, again, would have to be foreseen by the two or more persons concerned. The paradox still remains no matter how one attempt to twist or turn things around. *Unlimited foresight and economic equilibrium are thus irreconcilable with one another.* (Morgenstern 1935, p. 174)

In hindsight, this was really a truly important question that was raised by young and ambitious economist Morgenstern. Unfortunately, it remained long as a mere question. For a perfect solution for getting out of the paradox, Morgenstern at Vienna needed a mathematical help from von Neumann at Princeton. Even after the collaboration between Morgenstern and von Neumann began at Princeton, the paradox aforementioned gave them a constant headache. And in their joint work

Theory of Games and Economic Behavior, the Holmes-versus-Moriarty duel was once again taken up and very carefully investigated in the following way:

> The game to be considered below is an episode from the Adventure of Sherlock Holmes. Sherlock Holmes desires to proceed from London to Dover and hence to the Continent in order to escape from Professor Moriarty who pursues him. Having boarded the train he observes, as the train pulls out, the appearance of Professor Moriarty on the platform. Sherlock Holmes takes it for granted and in this he is assumed to be fully justified that his adversary, who has seen him, might secure a special train and overtake him. Sherlock Holmes is faced with the alternative of going to Dover or leaving the train at Canterbury, the only intermediate station. His adversary whose intelligence is assumed to be fully adequate to visualize these possibilities has the same choice. Both opponents must choose the place of their detrainment in ignorance of the other's corresponding decision. If, as a result of these measures, they should find themselves, *in fine*, on the same platform, Sherlock Holmes may with certainty expect to be killed by Moriarty. If Sherlock Holmes reaches Dover unharmed, he can make good his escape. (von Neumann and Morgenstern 1944, pp. 175–176)

The question of much interest is what are the good strategies for Sherlock Holmes and also what are the good strategies for Professor Moriarty. As is seen in Fig. 1.1, Holmes, who is leaving London toward Dover, has to choose one out of the following two strategies:

【Strategy D】Directly going to Dover, the destination.

【Strategy C】Quitting at Canterbury, only the intermediate station between London and Dover.

Moriarty, who is going in hot pursuit of Holmes, is also supposed to choose one out of the two strategies aforementioned; namely, Strategy D and Strategy C. Consequently, we should have the following four possible combinations of the two players:

【Combination of D and D】The two players meet at Dover. Holmes will certainly be killed by Moriarty.

【Combination of C and C】The two players meet at Canterbury. Holmes will certainly be killed by Moriarty.

【Combination of D and C】They do NOT meet at all. Holmes will quit at Canterbury and safely stay in England.

		Sherlock Holmes	
		Going to Dover	**Quitting at Canterbury**
Professor Moriarty	**Going to Dover**	They meet at Dover	They do NOT meet Holmes stays in Engl. **【Doyle's Story】**
	Going to Canterbury	They do NOT meet Holmes goes to the Continent	They meet at Canterbury

Fig. 1.1 Showdown between Professor Moriarty and Sherlock Holmes: *The Final Problem* in Conan Doyle (1894)

【Combination of C and D】They do NOT meet at all. Holmes will reach Dover unharmed, and make good his escape to the Continent.

Figure 1.1 is supposed to summarize the whole picture of *The Final Problem*, in which Moriarty and Holmes are the two key players and each one has the two strategies "Going to Dover" and "Quitting at Canterbury." According to Conan Doyle's original story, the "combination of D and C," implying that Moriarty and Holmes do not meet on the same platform and Holmes safely stays in England, should be the natural consequence of the situation. The question that would immediately occur is whether and to what extent Doyle's solution is justifiable. Is the solution "logically justifiable," and to what extent Doyle's solution is justifiable? Is the solution "logically justifiable," or "emotionally justifiable," or else?

At first glance, it looks like a game-theoretical situation. Even if we boldly attempt to confuse "the real world where flesh-and-blood persons emotionally interact with each other" with "the hypothetical world where robot-like persons exchange mechanical responses," however, we must take every possible care. Otherwise, a sort of Frankenstein's monster would unfold before our eyes. To this end, we have to make clear the following problems.

1. The first problem to clarify is concerned with the one of how much Holmes' intellectual faculties are in comparison with Moriarty's. On the one hand, it is true that Holmes always devotes himself to chemical experiments and shows flashes of genius in his reasoning as a detective. On the other hand, he is not only a heavy smoker and has little experience of having loved someone, but his criminal investigation has not been 100% perfect. For instance, in Doyle's first short story *A Scandal* in *Bohemia*, Holmes' skillful project was doomed to failure by the flash of wit of Irene Adler, who earned Holmes' unbounded admiration.[9]

   Professor Moriarty was a mathematical genius, and his career had been an extraordinary one. At the age of 21, he wrote an excellent treatise on the Binominal Theorem and won the Mathematical Chair at a university. But alas, he also had hereditary tendencies: a criminal strain ran in his blood, which was rendered more dangerous by his extraordinary mental powers. According to Sherlock Holmes, Moriarty was now "the Napoleon of crime," and the two men were presumably almost equally intelligent. We must say, however, that Moriarty's crime records had never been perfect, for some of his plans had to suffer considerable setbacks.[10]

---

[9] Although Sherlock Holmes is a brilliant detective, he is no more than a human being, sometimes making some human mistakes. In the story *A Scandal in Bohemia*, a lovely and cool female, called Irene Adler, is one of those women who have cleverly escaped from the pursuit of Holmes and London Police.

[10] According to Conan Doyle's detective story, "Monna Lisa," the famous masterpiece of Leonardo da Vinci, was once stolen from the Musée du Louvre (English: Louvre Museum). Sherlock Holmes, who was secretly asked to do criminal investigation from the French Government, suspected that Professor Moriarty and his criminal gang were working behind the scene and succeeded in getting back the masterpiece to the Museum. Such an outstanding achievement by Holmes, however, meant a devastating damage to Moriarty, who was then planning to kill Holmes, "the unbearable person."

For instance, against Holmes' repeated interferences, Moriarty made the following strong protest:

> 'You [Holmes] crossed my [Moriarty's] patch on the 4th of January,' said he [Moriarty]. 'On the 23rd you incommoded me; by the middle of February I was seriously inconvenienced by you; at the end March you I was absolutely Hampered in my plans; and now, at the close of April. I find myself placed in such a position through your continual persecution that I am in positive danger of losing my liberty. The situation is becoming an impossible one.' (Doyle 1894, pp. 336–337)

In short, no matter how clever Holmes and Moriarty might be, they should never be equivalent to Almighty God. Even if Holmes might think of Moriarty as a person of almost equivalent intelligence, it would be unlikely that Moriarty's intelligence is exactly as the same as Holmes' intelligence. Objectively speaking, the intelligence difference between the two persons might be very small. Subjectively, however, he is likely to believe that there should exist a non-ignorable difference. Holmes must be such a self-assertive person that he could induce the reader to believe in Holmes' final victory over Moriarty.

2. The second point is related to the physical strength of Holmes compared with Moriarty. When Holmes happens to fight hand to hand with Moriarty, the former who is a master of Japanese martial art must have an advantage over the latter. In fact, as is seen in Doyle's subsequent story *The Adventure of the Empty House* (Doyle 1903), Holmes takes such a technical advantage that he can knock down Moriarty into the Reichenbach Falls. To tell the truth, however, Moriarty has his own advantage over Holmes in that the remaining gang supporting the former may be strong enough to overpower the latter. Taking all those points into consideration, there would be no telling what will eventually happen between the two persons.

   Now let us take a closer look again at Fig. 1.1. Then, we immediately see the possibility that both Moriarty and Holmes meet face to face either at Dover or at Canterbury. If such a face-to-face meeting on the same platform happens, then it is conjectured by von Neumann and Morgenstern that "Holmes may with certainty expect to be killed by Moriarty" (see von Neumann and Morgenstern 1944, p. 177). We do not quite agree with this opinion, however. Generally speaking, assuming the defensive is a harder job than assuming the offensive. Sometimes, however, counterattacking an offensive enemy might be very effective. So, Moriarty's assault on Holmes might be successful with high probability, but not with 100% certainty. Or possibly, Holmes' counterattack on Moriarty could produce an unexpectedly good result. No one could tell us in advance which one is more likely than the other.

3. The third point involves the realistic validity of the probabilistic combination of any two strategies. Given the matrix of strategies in Fig. 1.1, we have to find "the optimal strategy" of the two players. In this regard, there is a critical gap between Doyle's *practical solution* and von Neumann–Morgenstern's *speculative solution*.

According to Doyle's real story *The Final Problem*, the combination of D and C represents the one of the optimal strategies of the two players. Holmes who has just quit at Canterbury is able to see Moriarty's special train passing at Canterbury toward Dover. Then, Holmes let a smile of triumph flash across his face, saying the following:

> 'There he goes,' said Holmes, as we watched the carriage swing and rock over the point. 'There are limits, you see, to our friend's intelligence. It would have been a coup-de-maître had he deduced what I would deduce and acted accordingly.' (Doyle 1894, p. 340)

We believe that this passage well describes Holmes' character as a human being and the possible limits of Moriarty's intelligence. It is quite remarkable to see, however, that Morgenstern raised a strong objection against Doyle's practical solution as follows:

> The narrative of Conan Doyle excusably disregards *mixed* strategies and states instead the actual developments. According to these, Sherlock Holmes gets out at the intermediate station and triumphantly watches Moriarty's special train going on to Dover. Conan Doyle's solution is the best possible under his limitations (to pure strategies), insofar as he attributes to each opponent the course which we found to be the more probable one (i.e. he replaces 60% probability by certainty). It is, however, somewhat misleading that this procedure leads to Sherlock Holmes's Complete victory, whereas, as we saw above, the odds (i.e. the value of a play) are definitely in favor of Moriarty. (Our result for ε, η yields that Sherlock Holmes is as good as 48% dead when his train pulls out from Victoria Station. Compare in this connection the suggestion in *Morgenstern*, loc. cit., p. 98, that the whole trip is unnecessary because the loser could be determined before the start.) (von Neumann and Morgenstern 1944, p. 178, footnote 1)

Mitsuo Suzuki (1959) was in strong support of von Neumann and Morgenstern (1944), voicing his own opinion as follows:

> The conclusion would state that Moriarty's best strategy should be going on to Dover with probability 3/5, and Holmes's best strategy, with probability 2/5. In other words, what Moriarty should do at this stage is to make the lottery which contains five tickets, with three for "going on to Dover" and two for "quitting at Canterbury," and then to decide which option to choose on the basis of the result of the lottery. (Suzuki 1959, p. 55)

In the above quotation, we note that the probabilities "3/5 (60%)" and "2/5 (40%)" are referred. The reason why those specific probabilities matter will be explained in detail later. Although this may be somewhat important, we do believe that there exists a more critical issue to discussed here. The key question to ask is related to the realistic validity of the concept of "mixed strategies" in the "once and for all game." In "The Final Problem," can we possibly imagine that Moriarty, who is desperately pursuing Holmes, quickly makes the relevant lottery with five tickets and simply decides his strategy by drawing lots? Consider the moment at which Holmes and Moriarty are just going for a one-on-one fight. If they have to decide their next strategies by drawing lots, then the duel scene will be greatly spoiled. No doubt, Holmes as a cool detective should instead adopt a "pure strategy" against Moriarty, a notorious villain.

The critical difference between "once and for all games" and "repeated games" must be recognized by any decision makers. We live in the world of uncertainty.

What may happen next, nobody can tell. People may not be informed of the exact shape of the distribution, but they have to make decisions on the basis of partial information. Under the conditions of "true uncertainty" a la Frank Knight (1921), we cannot depend on strict mathematical expectations since the basis per se for making such calculations no longer exists. In this connection, great economist John M. Keynes (1936) once made the following remark:

> Most, probably, of our decisions to do something positive, the full consequences of which will be drawn out over many days to come, can only be taken as a result of animal spirits of a spontaneous urge to action rather than inaction, and not as the outcome of a weighted average of quantitative benefits multiplied by quantitative probabilities. (Keynes 1936, p. 161)

If we take those things into consideration, we do not fully endorse the objections of von Neumann and Morgenstern (1944) and Suzuki (1959), but we feel ourselves rather attracted to Doyle's original solution. In short, we are not in a position to imagine the situation under which a gifted person like Holmes or Moriarty is drawing lots in his decision-making. Any brave man would have animal spirits, always taking a positive attitude and venturing his life.[11]

1.4 The Adventure of the Empty House: Sherlock Holmes Versus Colonel Moran

As was mentioned above, *The Final Problem* (1894) was intended to be exactly what its name meant. Sherlock Holmes struggled with Professor Moriarty over the Reichenbach Falls, with the result that the two together would fall into the falls. No falling body would possibly escape from inevitable death.

To Doyle's delightful surprise, however, something extraordinary happened. The public reaction to the apparent death of Sherlock Holmes was so overwhelming that Doyle was pressured into Holmes' miraculous revival in another story *The Adventure of the Empty House* (1903). Contrary to what Watson believed, Holmes definitely won against Professor Moriarty at Reichenbach Falls with the help of Japanese marshal art. Unfortunately, some of Moriarty's confederates knew that Holmes was still alive and tried to take every opportunity to kill Holmes. Colonel Sebastian Moran, nicknamed "the second most dangerous man in London" by Holmes, was presumably the most risky confederate as a trained sniper with a special air gun.[12]

[11] In recent times, "animal spirits" a la John M. Keynes and "courage to deal with uncertainty" a la Frank H. Knight have attracted the attention of economists again. For this point, see Sakai (2000).

[12] In my opinion, Colonel Moran plays an important role in Conan Doyle's detective stories. Moran is not a superman like Sherlock Holmes or Professor Moriarty, but very good at air rifle shooting. Although the importance of the duel between Moran and Holmes has been underestimated in the academia, we should pay more attention to it. In our opinion, another duel between Professor

		Sherlock Holmes	
		(Waiting at) the old apartment 【Moran's prediction】	(Waiting at) the empty house
Colonel Moran	(Shooting from) the outside 【Holmes's prediction】	Moran predicts Holmes predicts	Moran does NOT predict Holmes predicts
	(Shooting from) the empty house	Moran predicts Holmes does NOT predict	NEITHER Moran NOR Holmes predicts 【Doyle's Story】

Fig. 1.2 Showdown between Colonel Moran and Sherlock Holmes: *The Adventure of the Empty House* in Conan Doyle (1903)

One day, Holmes told Watson that they were going to do some dangerous work. They entered an empty house, an abandoned building whose front room overlooked Holmes' room, Baker Street. Very surprisingly, Holmes could be seen silhouetted against the blind: it was a waxwork bust, moved regularly by Mrs. Hudson to simulate life. Holmes used the dummy, anticipating an attempt on his life that very night. Holmes and Watson waited several hours in the empty house. A sniper fired an air gun to assassinate his foe. To Holmes' surprise, Moran as the sniper chose the empty house as his vantage point. Then, Moran was arrested and disarmed by Inspector Lestrade, the policeman asked to come by Holmes.

Doyle's new story *The Adventure of the Empty House* can be regarded as another masterpiece. Here, we can see a new sort of duel between Colonel Moran and Sherlock Homes. At first glance, this *new duel* seems to similar to the *old duel* between Professor Moriarty and Sherlock Holmes, but these two duels must be entirely different from an intellectual or informational viewpoint. Whereas Moran is a single-minded colonel with imperfect information, Holmes is a very intellectual detective with accurate information. Such a personal difference between the two persons would make a unique game play worthy of close investigation. As is seen in Fig. 1.2, Moran, who is anxious to kill Holmes by air gun, has to choose one out of the following two strategies:

[Strategy OU] Shooting from the outside **[Holmes' prediction]**

[Strategy EH] Shooting from the empty house

Holmes, who is threatened to be shot by an air gun, has to choose the following two strategies:

[Strategy OA] Waiting at the old apartment **[Moran's prediction]**

[Strategy EH] Waiting at the empty house

Moriarty and Sherlock Holmes should not be overestimated. A balanced view of the two duels would be required for full understanding of Holmes drama.

Given those strategies, namely two strategies for each player, we must have the following four possible combinations of the two players:

[Combination of OU and OA] This is both predicted by Moran and Holmes

[Combination of EH and EH] This is neither predicted by Moran and Holmes **[Doyle's own story]**

[Combination of OU and EH] This is NOT predicted by Moran, but surely predicted by Holmes

[Combination of EH and OA] This is predicted by Moran, but NOT so by Holmes

Figure 1.2 may summarize the whole picture of *The Adventure of the Empty House*. Remarkably, to our surprise, both Moran's prediction and Holmes' prediction are *not quite correct*! When Moran, a simple-minded person, saw a Holmes-like dummy silhouetted against the blind in the old apartment, he easily misunderstood that the silhouette represented the real Holmes. According to Moran's prediction, however, Holmes should be staying in the old apartment, but not in the empty house. In contrast to Moran, Holmes was a famous detective with high intelligence, so that he rightly predicted that Moran would shoot him by air gun. Holmes' prediction, however, was not quite correct: in fact, Moran attempted to shoot Holmes from the empty house, *not from the outside*. In reality, no man is perfect; even Holmes made a mistake.

We admit that there have been so many Holmes fans around the world. They are constantly reading those Doyle's detective stories, which include the first story *The Final Problem* and the second story *The Adventure of the Empty House*. Although the first story looks more famous than the second one, we believe that the two stories require equal treatment for scientific investigation.

1.5 Games Under Risk Versus Games Under Uncertainty

As was noticed at the beginning, there are two memorable years for the theory of games: 1928 as the Year of Birth and 1944 as the Year of Establishment. So, to believe or not, the centenary anniversary of game theory is steadily approaching.

We must bear in mind that game theory is not merely a brainchild of mathematical genius John von Neumann, but rather a product of compassionate economist Oskar Morgenstern as well. Since Morgenstern was a passionate Sherlock Holmes fan, Conan Doyle's detective story, featuring Holmes as its main actor, has produced a far-reaching effect on their joint product *Theory of Games and Economic Behavior* (1944). We must not forget the fact that from the start, the two fantastic stories *The Final Problem* (1894) and *The Adventure of the Empty House* (1903) were destined to deal with different sorts of game theoretic situations.

In *The Final Problem*, there are the outstanding two players, i.e., famous detective Sherlock Holmes and notorious mathematician Professor Moriarty. Each player, who lives up to his reputation as a genius in his own field respects each other, can perfectly foresee the *set* of possible strategies of his opponent. It is true that he has no power to exactly anticipate which action his enemy really takes. Whichever action the enemy takes, however, it should be only *one element* of the whole anticipated set, so that there would be no surprises whatever. This is no less than the world of games under *risk*, with no *black swans* a la Taleb (2007) being present.

Turning to another story *The Adventure of the Empty House*, we get into the new world of unexpected results and surprises. Here again, there are the two players, namely super detective Sherlock Holmes and simple-minded Colonel Moran. Although their intelligence levels are poles apart, they are just human beings who are apt to make some mistakes. Although Holmes is thought of as a man of super brain, he should be no way free of making mistakes. Here, each player cannot even foresee the set of possible strategies of his opponent, so that unexpected events, which do not belong to the anticipated set, are likely to take place. This is really the world of games under *true uncertainty*, in which the concept per se of an equilibrium of the game ceases to be effective.[13]

Now, if we look hard at the real world today, it should be worth questioning which world we live in, *the World under Risk* or *the World under Uncertainty*. We do think that the real events we observe in our daily lives bear more resemblance to the second story *The Adventure of the Empty House* than to the first story *The Final Problem*.

Close examination of the real world may tell us that we live in the World of Uncertainty in which everything including human beings is uncertain and unreliable. It is true that some persons are very clever and resourceful. Cleverness and rich resource, however, do not guarantee correct judgment and perfect forecasting. Besides, we should not forget the sad fact that both intelligence and information are not fairly distributed among the people. It seems that the majority of the people behave like mediocre Colonel Moran, neither like superman Sherlock Holmes nor like Professor Moriarty.

In conclusion, it is high time for us to go beyond the "clean and beautiful world of von Neumann and Morgenstern." We have to find a better way, not even the best way of life, in such an "unclean and dirty world" as we see it.

References

Aumann RJ, Hart S (eds) (1992) Handbook of game theory and economic applications. North Holland

[13] According to Frank H. Knight (1921), "non-measurable uncertainty" should be distinguished from "measurable risk." In the measurable world, while risk is a valid concept, uncertainty tends to lose its validity. We believe that there is an incurable gap between game theory and uncertainty theory. It is high time to establish a *new* theory of unexpected surprises, going far beyond the narrowed scope of *old* theory of games. Also see Kuhn and Nasar (2002), Nasar (1998), Sakai (1982, 1996, 2000, 2005, 2010, 2019), Soros (2008), and Thurow (1980).

Dasgupta P, Gale D, Hart O, Maskin E (eds) (1992) Economic analysis of markets and games. The MIT Press, Cambridge, MA
Dixit AK, Nalebuff BJ (1991) Thinking strategically: the competitive edge in business, politics and economic life. Norton, New York
Doyle AC (1894) The final problem. In: The memoirs of Sherlock Holmes. George Newnes, London
Doyle AC (1903) The adventure of the empty house. In: The return of Sherlock Holmes (1905). George Newnes, London
Gamow G (1947, rev. 1961) One two three ... infinity. Viking Press
Keynes JM (1936) The general theory of employment, interest and money. Macmillan, London
Knight HF (1921) Risk, uncertainty and profit. University of Chicago Press
Kuhn HW, Nasar S (eds) (2002) The essential John Nash. Princeton University Press
Luce RD, Raiffa H (1957, rev. 1989) Games and decisions: introduction and critical survey. Wiley, New York
Marx K, Engels F (1848) Manifest der Kommunistischen Partei (English: The manifesto of the communist party). A pamphlet. London
Morgenstern O (1928) Wirtschaftsprognose, einer Untersuchung ihrer Voraussetzungen und Mölichkeiten (English translation: Economic forecasting: a study of its difficulties and impossibilities). Springer, Vienna
Morgenstern O (1935) Vollkommene Voraussicht und wirtschaftliche Gleichgewicht (English: Perfect foresight and economic equilibrium). Z National 6(3):337–357
Morgenstern O (1958) John von Neumann, 1883–1957. Econ J 68:499–503
Morgenstern O (1972) Thirteen critical points in contemporary economic theory: an interpretation. J Econ Lit 10(4):1163–1189
Morgenstern O (1976) The collaboration between Oskar Morgenstern and John von Neumann. J Econ Lit 14(3):805–816
Nakayama C (1997) The process of collaboration between von Neumann and Morgenstern. History Econ Rev 26:40–50
Nakayama M, Muto S, Funaki Y (eds) (2000) Thinking in Game theory. Keiso Publishers, Tokyo
Nasar S (1998) A beautiful mind: the life of mathematical genius and Nobel laureate John Nash. Touchstone, New York
Okada A (1996, rev. 2011) Game theory. Yuhikaku Publishers, Tokyo
Poundstone W (1992) Prisoner's dilemma: John von Neumann, game theory, and the puzzle of the bomb. Doubleday, New York
Sakai Y (1982) The economics of uncertainty (in Japanese). Yuhikaku Publishers, Tokyo
Sakai Y (1996) The economics of risk: information and culture (in Japanese). Yuhikaku Publishers, Tokyo
Sakai Y (2000) The concept of risk in economics: the impact of sympathy and animal spirits (in Japanese). Economic Planning Agency, Tokyo
Sakai Y (2001) Von Neumann, Morgenstern, and the world of game theory: part 1 (in Japanese). Tsukuba Econ Stud, Spring: 1–42
Sakai Y (2005) On risk society (in Japanese). Iwanami Publishers, Tokyo
Sakai Y (2010) The economic thought of risk and uncertainty (in Japanese). Minerva Publishers, Kyoto
Sakai Y (2019) J.M. Keynes versus F.H. Knight: risk, probability, and uncertainty. Springer Nature, Singapore
Schotter A (ed) (1976) Selected economic writings of Oskar Morgenstern. New York University Press, New York
Shubik M (1967) Essays in mathematical economics: in honor of Oskar Morgenstern. Princeton University Press
Shubik M (1982) Game theory in the social sciences: concepts and solutions. The MIT Press, Cambridge, MA
Soros G (2008) The crash of 2008 and what it means. Public Affairs, New York

Suzuki M (1959) The theory of games (in Japanese). Keiso Publishers, Tokyo

Suzuki M (1994) The new theory of games (in Japanese). Keiso Publishers, Tokyo

Suzuki M (1999) The world of game theory (in Japanese). Keiso Publishers, Tokyo

Taleb NN (2007) The black swan: the impact of the highly improbable. Random House, New York

Thurow LC (1980) The zero-sum society. Basic Books, New York

Varoufakis Y (ed) (2001) Game theory: critical concepts in the social sciences. Routledge, London

von Neumann J (1928) Zur Theorie der Gesellschaftsspiele (English translation: The theory of social game). Math Ann 100:295–320

von Neumann J, Morgenstern O (1944) Theory of games and economic behavior. 2nd edn, 1946; 3rd edn, 1953. Princeton University Press

Yoshioka S (1996) Poe's short stories and mathematics (in Japanese). In: Iyanaga S et al (eds) Mathematics and human life. Gakuseisha Publishers, Tokyo

Chapter 2
Reassessing Zero-Sum Games: Various Types of Matching Pennies

Abstract This chapter continues to critically reassess the significance and limitations of zero-sum games, which were jointly established by mathematician John von Neumann and economist Oscar Morgenstern during the difficult times of Second World War. First of all, we will intensively discuss several Zero-Sum Two-Person Games, with special reference to the Games of Matching Pennies. New graphical illustrations will be attempted for clarification of the matter. Then, we will turn to novelist Edgar Alan Poe's best story *The Purloined Letter*, showing how the Poe story may challenge the validity of Neumann–Morgenstern approach. Next, we will shed new light on Conan Doyle's famous story *The Final Case* as a variant of the Games of Matching Pennies. Finally, we will pick up the Game of Stone, Paper, Scissors, which constitutes another important member of Zero-Sum Two-Person Games.

Keywords Zero-sum two-person games · Graphical illustrations · The game of matching pennies · Poe's challenge · Doyle's story *The Final Case* · The game of stone, paper, scissors

2.1 Historical Background of the Theory of Games

The Year of 1944—this is certainly the most memorial year of the history of the theory of games. It is in this year that the collaboration between mathematician von Neumann and economist Oskar Morgenstern successfully produced the great book *Theory of Games and Economic Behavior* at the Princeton University Press.

In historical perspective, the devastating Second World War began in 1939 and finally ended in 1945. So, we could say that presumably, game theory was an academic by-product of the global war involving so many countries across the

---

This chapter is based on Sakai (2001), with many extensions and new figures. I am grateful to James Friedman (Rochester), Edward Zabel (Rochester), Alvin Roth (Pittsburgh), Richard Thaler (Chicago), Mamoru Kaneko (Tsukuba), and Mitsuo Suzuki (Tokyo Kogyo) for valuable suggestions.

Y. Sakai, *Games, Decisions, and Markets*, New Frontiers in Regional Science: Asian Perspectives 74, https://doi.org/10.1007/978-981-99-5285-4_2

Atlantic, the Indian, and the Pacific Oceans. Yasuhiro Sakai, the author of this book, was born in Japan 1 year after the Poland Invasion by Adolf Hitler and 4 years before the publication of *Game Theory* by Neumann and Morgenstern. Let us recall the political and economic situations surrounding Japan in 1944. It is true that at the beginning of the Pacific War with the United States, the Empire of Japan won a number of battles after it made a surprise attack on the Pearl Harbor in December 8, 1945 (Asia/West-Pacific time zones). As the saying goes, however, happy events are often accompanied by difficulties.

We can recall what Admiral Isoroku Yamamoto (1884–1943), a legendary Japanese tragic hero, told his fellow career soldiers just before the Pacific War began, "In the first six to twelve months of the war with the United States and Great Britain, I will run wild and win victory upon victory. But then, if the war protracts two or three years, I have no expectation of success." In retrospect, Yamamoto was quite right, and his prediction came true. After so many Japanese aircraft carriers were destroyed at the notorious Midway Battle in June 1942, the Japanese navy was doomed to total defeat in power struggles against the American counterpart, and even Admiral Yamamoto himself was killed in action in the South Pacific in 1943.

Under those difficult situations under the war period, all the scholars around the world had to keep trying their best in order to pull through the crisis. In Japan, which engaged in the war against the United States, there were a selected number of people who made every possible effort to continue their research activities. Among those devoted people was Professor Yasuma Takata (1883–1972) of Kyoto Imperial University, who as an influential economist and sociologist made a great contribution to the Power Theory of Social Science. His productive power as a serious scholar during the hectic war period was exceptionally great, for he published so many books on social power, race and economy, interest theory, controlled economy, and many other topics as well illustrated by Takata (1937, 1940a, b, 1941, 1942, 1943, 1944). In spite of the decline of a nation as a whole, Takata's personal spirits were still so high that he bravely declared as follows. "There might be a sudden collapse in the fortunes of the country." Takata is nevertheless ready to stand up alone against the stream.[1]

When the United States became a formidable enemy of Japan, it embarked on a new project, called the Manhattan Project, whose main purpose was to produce weapons of mass destruction including atomic bombs. On the one hand, von Neumann himself gave everything he acquired in order to invent a very powerful computer that was necessary to implement the project, thus putting less emphasis on the project itself. On the other hand, von Neumann' s commitment to Game Theory had to become less and less as the war went on. Nevertheless, his collaboration with

[1] Yasuma Takata (1883–1972), called "Marshall of Japan" by Martin Bronfenbrenner, has made an outstanding contribution to the "Theory of Social Power." For the establishment of a New Theory of Social Science, we have to combine power theory a la Takata and game theory a la Neumann and Morgenstern. As the saying goes, where there is a will, there is a way! Also see Takata (1955) and Morishima (1994; 1998).

economist Morgenstern somehow continued at Princeton until their bulky book *Theory of Games and Economic Behavior* was finally finished in 1944. In short, game theory was truly the joint product of mathematical talent of von Neumann and humanist intelligence of Morgenstern.[2]

As the reader has noticed, Chap. 1 and this chapter (i.e. Chap. 2) are a sort of "twin chapters." Taken together, those chapters are intended to critically reassess the significance and limitations of Zero-Sum Games, which were jointly established by mathematician John von Neumann and economist Oskar Morgenstern during the difficult times of Second World War.

Let us write down the contents of this chapter. The next section will carefully examine how their joint book has been organized, thus unveiling the real intention of the two authors. Section 2.3 will intensively discuss several zero-sum two-person games, with special reference to the Games of Matching Pennies. New graphical illustrations will be attempted for clarification of the matter. In Sect. 2.4, we will turn to novelist Edgar Alan Poe's best story *The Purloined Letter*, discussing how the Poe story may challenge the validity of Neumann–Morgenstern approach to human behavior. We note that this would be probably the first time to connect the creative writer Poe and the strong Neumann–Morgenstern alliance. In Sect. 2.5, we are in a position to safely return to our favorite story *The Final Problem* as a variant of the Game of Matching Pennies, which has continuously been our major topic in the last chapter, namely Chap. 1. In Sect. 2.6, we will pick up the Game of Stone, Paper, and Scissors, which clearly constitutes another important member of the family of Zero-Sum Two-Person Games. And final remarks will be made in Sect. 2.7.

2.2 The Great Book *Theory of Games and Economic Behavior*: Its Contents Critically Examined

We are now ready to critically examine the whole contents of the great book *Theory of Games and Economic Behavior* (1944) from a set of fresh angles. Honestly speaking, it does not look like a "real thing," but rather like an "imaginary creature." It is really a bulky book containing 600 and more pages, full of strange notations and pretentious equations. In today's house building, one of the most popular styles is a "barrier-free style," in the sense that many walking obstacles in the house should be avoided. Contrary to such a modern style, there exist so many obstacles and barriers

[2]It is said that human beings have three different mental temperaments, namely "intelligence," "passion," and "determination." While Morgenstern is dominantly "the man of strong passion," von Neumann is "the man of cool intelligence." The collaboration between the two persons seems to have yielded "solid determination." It is a sad historical fact that von Neumann, cool-headedly calculating the cost and balance of any project, once recommended the American Government to drop an atomic bomb to Kyoto, the ancient capital of Japan with so many temples and memorials. Unfortunately, there seemed to exist no upper bounds for cool calculation. For details, see Poundstone (1992).

Table 2.1 *Theory of Games and Economic Behavior*: Zero-Sum Games representing the Main Core

Contents	Chapter	Pages	Title
General Intro	1	1–45	Formation of the Economic Problem
	2	46–84	General Description of Games of Strategy
Zero-Sum Games	3	85–168	Zero-Sum Two-Person Games: Theory
	4	169–219	Zero-Sum Two-Person Games: Examples
	5	220–237	Zero-Sum Three-Person Games
	6	238–290	General Theory of Zero-Sum n-Person Games
	7	291–329	Zero-Sum Four-Person Games
	8	330–338	Some Remarks on $n \geqq 5$ Participants
	9	339–419	Composition and Decomposition of Games
	10	420–503	Simple Games
Non-Zero-Sum Games	11	504–586	General Non-Zero-Sum Games
	12	587–616	Concepts of Domination and Solution

in the game-theory architecture made by von Neumann and Morgenstern. We should stress to say, however, that it is no less than a kind of "Bible of Game Theory," which is worthy of attempting to overcome mathematical obstacles/barriers.

In the following, let us more carefully shed new light on the original book of game theory. As the saying goes, seeing is believing. Take a very close look at Table 2.1. Then, we will see that broadly speaking, it consists of three parts, namely General Intro, Zero-Sum Games, and Non-Zero-Sum Games.

First of all, we are most likely to be shocked to see the apparently strange compactness of "Preface to First Edition," which was written in January 1943. Contrary to our expectation, this important preface was very compactly written, thus being unduly condensed into following two paragraphs:

> This book contains an exposition and various applications of a mathematical theory of games. The theory has been developed by one of us since 1928 and is now published for the first time in its entity. The applications are of two kinds: On the one hand to games in the proper sense, on the other hand to economic and sociological problems which, as we hope to show, are best approached from this direction. The applications which we shall make to games serve at least as much to corroborate the theory as to investigate these games. The nature of this reciprocal relationship will become clear as the investigation proceeds. Our main interest is, of course, in the economic and sociological direction. Here we can approach only the simplest questions. However, these questions are of a fundamental character. Furthermore, our aim is primarily to show that there is a rigorous approach to these subjects, involving, as they do, questions of parallel or opposite interest, perfect or imperfect information, free rational decision or chance influences. (von Neumann and Morgenstern 1944, Preface to the first edition)

Interestingly enough, such unduly compactness of Preface was not confined to the First Edition. For instance, "Preface to Second Edition" (written in September 1046) and "Preface to Third Edition" (written in January 1953) were respectively one-page long and two-page long. Honestly speaking, the authors had little interest in making the prefaces well appealing to the reader.

Secondly, in sharp contrast to the brief Preface, the Main Text was quite long and powerful. Metaphorically speaking, if we are allowed to compare the full contents of the book to the full course of Western-style cooking, then Preface might be regarded as a "light appetizer," but the Main Text must be a "heavy main dish," and the Math Appendix a "colorful dessert."

As can be seen in Table 2.1, the contents of the book are divided into three parts. The first part deals with General Intro, consisting of the first two chapters, namely Chaps. 1 and 2. The second part systematically discusses Zero-Sum Games, containing eight chapters from Chap. 3 through Chap. 10. The third part is concerned with the final two chapters, i.e., Chaps. 11 and 12. Easy calculation can tell us that the shares of Part I, Part II, and Part III are respectively 13.64%, 68.02%, and 18.34%. Therefore, Zero-Sum Games, and especially Zero-Sum Two-Person Games, have been the main target of the joint work of von Neumann and Morgenstern. This should clearly indicate the third characteristic of the 1944 book.

We wonder why the two giants of modern times had been so obsessive about zero-sum games. There must be some reasons for such obsession. The first reason we can think of seems to be historical. Let us recall the historical backgrounds of the two superstars. Both von Neumann and Oskar Morgenstern were born in Europe or the "Old World." While von Neumann spent his young days as the son of a rich Jewish family, he had to experience the suffering and tribulations of human life as the Nazis led by Adolf Hitler (1889–1945) began to assert itself in the region. Shortly after Morgenstern became a professor of economics at the prestigious University of Vienna, his chair was suddenly sacked because he was accused of being "politically unbearable." Although both von Neumann and Morgenstern later settled down at the Institute of Advanced Study, Princeton University in the United States, i.e., the "New World," their suffering and hardship in the "Old World" never escaped their memories. Besides, when the Polish Invasion by Dictator Adolf Hitler triggered the Second World War, the peaceful private lives of von Neumann and Morgenstern were no longer isolated from the busy public duties. In short, together with the general public, the two great men were also driven into a corner, having the mental state that "To fight or not, that is the question." We strongly believe that behind the birth and development of game theory, there existed a number of fierce rivalries between states, between classes, and between ideologies.

The second reason for the rise of game theory is not historic, but rather more technical. According to von Neumann and Morgenstern, zero-sum games and non-zero-sum games should not be so hostile as they appear. Indeed, if we additionally assume the existence of a "third hypothetical player," then it is theoretically feasible to *expand* non-zero-sum two-person games to Zero-Sum Three-Person Games. In other words, in general, it is technically possible to *embed non-zero-sum n-person games into Zero-Sum n + 1-Person Games*. In this connection, von Neumann and Morgenstern remarked the following:

> We shall primarily construct a theory of the zero-sum games, but it will be found possible to dispose, with its help of all games, without restriction. Precisely: We shall show that the general (hence in particular the variable sum) n-person game can be reduced to a zero-sum $n + 1$ person game. Now the theory of the zero-sum n-person game will be based on the special case of the zero-sum two-person game. Hence our discussion will begin with a theory of these games. (von Neumann and Morgenstern 1944, p. 47)

Remarkably, very similar remarks can be seen again much later in the book:

> Any given general games can be re-interpreted in a zero-sum game. This may seem paradoxical since the general games form a much more extensive family than the zero-sum games. However, our procedure will be to interpret an n-person general game as an $n + 1$-person zero-sum game. . . . The procedure by which a given general n-person game is re-interpreted as an $n + 1$-person zero-sum game is a very simple and natural one. It consists of introducing a fictitious—$n + 1$-st player who is assumed to lose the amount which the totality of the other n—real—players wins and vice versa. (von Neumann and Morgenstern 1944, pp. 505–506)

The operations of *expansion and embedding*, which should be very important, were favored and often used by von Neumann and Morgenstern. To our astonishment, however, these operations have been all but forgotten in the recent discussions on game theory. It is now high time to recover and further develop the skillful operations of expansion and embedding. As the saying goes, better late than never!

2.3 Zero-Sum Two-Person Games: Pure Strategies and Strictly Determined Games

In the Great Book *Theory of Games and Economic Behavior*, there is one important subject that plays the leading role throughput the book. That subject is nothing but the one named *Zero-Sum Two-Person Games.*

While we limit attention to those zero-sum two-person games, we have to keep in mind that there are two different kinds of games. The first kind is called *a strictly determined game or a closed game*, in which an equilibrium solution can be found in the specific set of *pure strategies.* The second kind is named a *non-strictly determined game or an open game*, where an equilibrium solution cannot be determined by the limited range of pure strategies, but should be found in the largely expanded set of *mixed strategies or the stochastic combinations of pure strategies.*[3]

[3]In the real world, we must notice that there is a *third kind of game* in the sense that it can be described neither by a closed game nor by an open game. von Neumann and Morgenstern (1944) already paid attention to "some games of peculiar character" as follows: "Some games like Roulette are of an even more peculiar character. In Roulette the mathematical expectation of the players is clearly negative. Thus the motives for participating in that game cannot be understood if one identifies the monetary return with utility" (von Neumann and Morgenstern 1944, p. 87, footnote 3). We remember that in the Pacific War, an apparently crazy "Kamikaze pilot" made a dash at an American battleship. The motive of the Kamikaze pilot for that action cannot be understood if his return is merely identified with money. Also see Morgenstern (1928, 1935, 1958, 1972, 1976),

		Player *B* Pure Strategies β_1	β_2	**Row Max (security level)**
Player *A*	α_1	a_{11}	a_{12}	**Min**$_j$ a_{1j} → v_A = **Max**$_i$ **Min**$_j$ a_{ij}
Pure Strategies	α_2	a_{21}	a_{22}	**Min**$_j$ a_{2j}
Column Max (security level)		Max$_i$ a_{i1}	Max$_i$ a_{i2}	
	→	v_B = **Min**$_j$ **Max**$_i$ a_{ij}		

Fig. 2.1 Zero-Sum Two-Person Game. When $v_A = v_B$, the common value shows the value of a strictly determined game

Frankly speaking, it is true that von Neumann and Morgenstern placed more emphasis on non-strictly determined games than strictly determined games. In order to climb up to the top of a high mountain peak, however, it would be a good idea to start with a sequence of lower hills and mountains. Likewise, we would like to begin with a group of simpler strictly determined games before challenging more general and more difficult (non-strictly determined) games.

In what follows, we would like to carefully discuss the basic structure of a zero-sum two-person game. To this end, let us suppose that the two players, Player A and Player B, engage in a certain game. As is seen in Fig. 2.1, whereas Player A has two pure strategies, α_1 and α_2, Player B has two strategies, β_1 and β_2. If A chooses α_i and B chooses β_j, then A's outcome is shown by a_{ij} $(i, j = 1, 2)$. Since this is a zero-sum game, the sum of A's outcome and B's outcome must be zero. This implies that under the present situation, B's outcome must be given by $(-a_{ij})$.

Von Neumann and Morgenstern assume that each player, Player A and Player B, is a very defensive person in the sense that he follows the rule of "safety first." We think that this would be a debatable assumption, perhaps provoking someone to raise objections. However, we will be content here to just obey the Neumann–Morgenstern assumption that each player always "play safe" in any possible situation.

When Player A chooses strategy α_1, which strategy, β_1 or β_2, A's opponent B chooses cannot be known to A in advance. All we can say at this point is that A's payoff is a_{11} if B chooses strategy β_1, and a_{12} if B chooses strategy β_2. Hence, whichever strategy B chooses, *at least* the amount of $\text{Min}_j a_{1j}$ is guaranteed as a "security level" by A's choice of α_1. Therefore, we may call $\text{Min}_j a_{1j}$ the *security level for Player A of strategy* α_1. In a similar fashion, the *security level for Player A of strategy* α_2 is shown by $\text{Min}_j a_{2j}$. The purpose of Player A's choice behavior is to choose a strategy that makes his security level as much as possible. In other words,

---

Morishima (1994), Luce and Raiffa (1957), Aumann and Hart (1992), Dasgupta et al. (1992), Dixit and Nalebuff (1991), Okada (1996, 2nd ed. 2011), Poundstone (1992), Sakai (1982, 1996, 2000, 2001, 2005, 2010, 2019), Schotter (1976), Shubik (1967, 1982), Suzuki (1959, 1994, 1999), Taleb (2007), and Varoufakis (2001).

Player A's behavior principle is summarized as the *max-min strategy*, and *the value of the game for Player A* is thus given by the following equation:

$$v_A = \mathrm{Max}_i\ \mathrm{Min}_j\ a_{ij}. \tag{2.1}$$

Now, we recall that the behavior rules of the two players, Player A and Player B, should be just symmetrical. For Player B, an analogous situation must obtain. It is reasonable to call $\max_j a_{ij}$ the security level of strategy β_j, for any loss greater than level is effectively avoided.

The purpose of Player B's choice behavior is to choose a strategy that makes his loss level as less as possible. In other words, Player A's behavior principle is summarized as the *min-max strategy*, and *the value of the game for Player B* is thus given by the following equation:

$$v_B = \mathrm{Min}_j\ \mathrm{Max}_i\ a_{ij}. \tag{2.2}$$

It is quite clear that in general, we have $v_A \geqq v_B$. In particular, if v_A and v_B happen to be just equal, then we say that *the solution of the game*, v, exists, thus being written as follows:

$$v = v_A = v_B. \tag{2.3}$$

Given any zero-sum two-person game, there is in general no guarantee that the solution of the game is provided by pure strategies only. The solution exists in some cases, but it does not exist in other cases. When we carefully read *Theory of Games and Economic Behavior* (1944), we find the interesting fact that there is only one solvable case, which has been dealt with by von Neumann and Morgenstern. Such an "exceptional case" is shown in Fig. 2.2.

Figure 2.2 consists of two charts, Chart (A) and Chart (B). First of all, by looking at Chart (A), we find "Max of Row Min $= -1$" on the right of the outcome matrix and "Min of Column Max $= -1$" on the bottom of the matrix. This common value, circled by a dotted line, must represent the value of the game; namely, $v = v_A = v_B = -1$. Besides, the corresponding payoff (-1) in the payoff matrix is vividly circled by a solid line.

Let us figure out a good way to handle the present situation. Our unique device is based on drawing a three-dimensional diagram as shown in Chart (B). Take a careful look at two group of directions; one group, containing α_1 direction and α_2 direction, going from the near side to the far side, and another group, consisting of β_1 direction and β_2 direction, going from left side to the right side. Then, interestingly enough, we will find that Point a_{21}, indicated by black ball •, is a *saddle point* in the sense that it is minimal in α-direction and maximal in β direction. This clearly teaches us that any solution of the game, v, may be characterized as a saddle point.

We wonder if there might exist some suspicious readers who cast skeptical eyes upon the view of regarding a_{21} as a saddle point. In Fig. 2.2, since all the four points,

(A) Pure strategies and payoff matrix

		Player *B*		
		β_1	β_2	Row Max
Player *A*	α_1	−2	1	−2
	α_2	[−1]	2	[−1] Max
Column Max		[−1] Min	2	

(B) Graphical representation of the original game

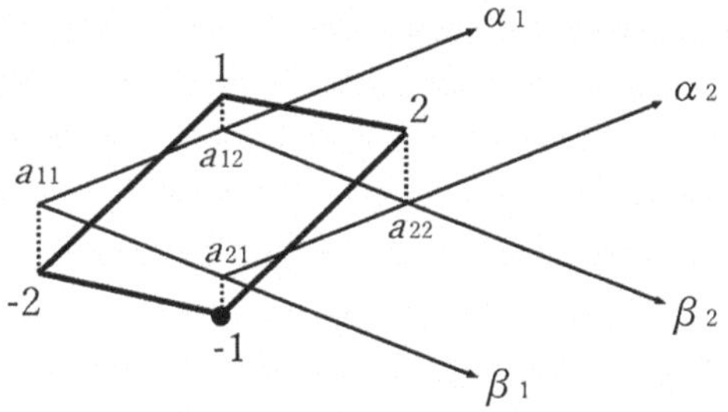

Fig. 2.2 A strictly determined game: the solution graphically illustrated

a_{11}, a_{12}, a_{21}, and a_{22}, are *end points*, the unique convex–concave shape of a saddle point is not well pictured in such a simple diagram. It is in this sense that a_{21} might not look like "a full-fledged saddle point" to some non-mathematical readers. In order to mend such imperfection, let us conveniently introduce "supplementary strategies" into Chart (B). Then, we will nicely find that Point a_{21}, indicated by black ball •, is a *saddle point* in the sense that it is minimal in "α-direction" and maximal in "β-direction." This clearly teaches us that any solution v of the game may be characterized as a saddle point.

More specifically, as is seen in Chart (A) of Fig. 2.3, we are ready to draw an "expanded outcome matrix" in which we introduce α_3 as a supplementary strategy of Player *A*, and β_0 as a supplementary strategy of Player *B*. Then, the reader will easily understand the great impact of supplementary strategies, for the new strategy α_3 is a strategy inferior to the old strategy α_1, and the new strategy β_0, inferior to the old strategy β_1. We should keep in mind that the "expanded game" added with supplementary strategies has essentially the same structure as the "original game." The graphical representation of such an expanded game is depicted in Chart (B) of Fig. 2.3. We are now happy to see that Point a_{21}, indicated by black ball •, is really a charming *saddle point* to everybody's eyes. Thus, introduction of the two newly added strategies, α_3 and β_0, is quite appealing to our eyes. As the saying goes, seeing is believing indeed!

(A) New strategies α_3 and β_0 added to the matrix

		Player *B*				
		β_0 (newly added)	β_1	β_2	**Row Min**	
Player *A*	α_1	−1	−2	1	−2	
	α_2	0	[−1]	2	[−1]	**Max**
(newly added)	α_3	−1	−2	0	−2	
Column Max		0	[−1] **Min**	0		

(B) The point a_{21} is a saddle point in the expanded game

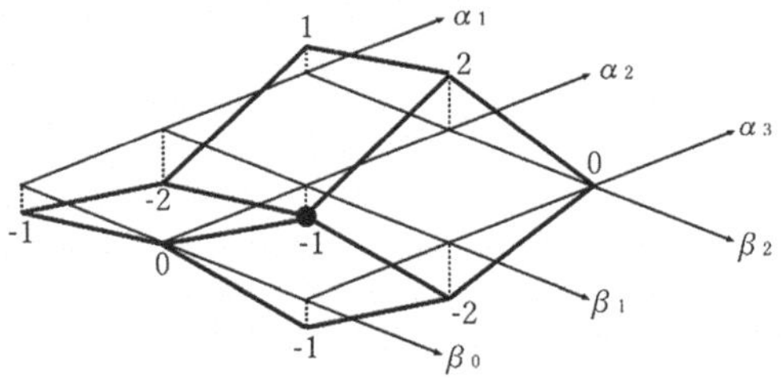

Fig. 2.3 The expanded game with newly added strategies, α_3 and β_0

2.4 The Game of Matching Pennies: Mixed Strategies and Non-strictly Determined Games

According to von Neumann and Morgenstern (1944), *zero-sum two-person games* are clearly the most important games. Remarkably, those games per se can be further divided into two different kinds. The first kind of games are classified as *strictly determined games*, whose solutions are determined by *pure strategies only*. Although those kinds of games have been exclusively discussed so far, however, they are relatively a few. In the real world, however, the second kind of games, or exactly *non-strictly determined games*, whose solutions should be written as *the mixtures of pure strategies*, or briefly *mixed strategies*, are more important.

Like many other scientists, von Neumann and Morgenstern got wild with excitement over many indoor games. Among those games, they had much interest in the following two games. They are the Game of Matching Pennies and the Game of Stone, Paper, Scissors. In fact, von Neumann and Morgenstern wrote the following impressive remark:

> In order to overcome the difficulties in the non-strictly determined case, it is best to reconsider the simplest examples of this phenomenon. These are games of Matching Pennies and of Stone, Paper, Scissors. Since an empirical, common-sense attitude with respect to the "problems" of these games exists, we may hope to get a clue for the solution of non-strictly determined (zero-sum two-person) games by observing and analyzing these attitudes. (von Neumann and Morgenstern 1944, p. 143)

Honestly speaking, the great book *Theory of Games and Economic Behavior* is not an easy book to read. First, it is a bulky book containing 600 and more pages, written in apparently German-like complicated English. Second, it looks like a Book of Applied Mathematics, with heavy mathematical equipment. There are many strange notations, overwhelming figures and involved computations. Third, it is not so well organized as it should be. Unnecessary repetition of similar, and even the same, topics can be seen almost everywhere throughout the book. For instance, the two authors' pet subject "The Game of Matching Pennies" has been repeatedly referred to so many times; in page 94, page 111, pages 143–144, page 164, page 166, page 169, pages 175–178, page 185, and so on. As the saying goes, perseverance may accomplish many things. We do not quite understand, however, why very similar subjects are scattered in so many pages.

von Neumann and Morgenstern (1944) is generous enough to discuss four different cases for "The Game of Matching Pennies." While the first three cases will be dealt with in this section, the last fourth case, which is nothing but the story *The Final Problem* in Conan Doyle's *Memoirs of Sherlock Holmes* (1892), will very carefully be reinvestigated in the next section.

Let us take a close look at Fig. 2.4. It clearly shows *The First Game of Matching Pennies*, or *Matching Pennies in Ordinary Form.* In Chart (A), there are two players in the game, namely Player A and Player B. We assume that each player has two strategies "H" (head) and "T" (tail). On the one hand, if A's strategy and B's strategy are *just matching* like "*H versus H*," or "*T versus T*," then A gains the amount (+1) as the winner 's prize (but B suffers the amount (−1) as the loser's loss). On the other hand, if the two persons' strategies are *non-matching* like "*H versus T*," *or* "*T versus H*," then A suffers the amount (−1) as the loser's loss (but B gains the amount (+1) as the winner's prize). As Fig. 2.4 tells us, the relevant outcome matrix becomes a symmetric matrix with all the diagonal elements (+1) and all the off-diagonal elements (−1).[4]

Let us calculate the values of the game for Players A and B. Then, we will easily find the following:

$$v_A = \mathrm{Max}_i \ \mathrm{Min}_j \ a_{ij} = -1, \tag{2.4}$$

[4] Among those four cases of Matching Pennies, the first case was especially referred to as "Matching Pennies in its ordinary form." The remaining three cases were discussed as "Matching Pennies in other forms." For details, see von Neumann and Morgenstern (1944), pp. 175–178.

(A) Payoff matrix: limiting pure strategies does not yield solutions

		Player *B*				
		H	*T*	**Row Min**		***DIFFERENCE*** (*optimum ratio*)
Player *A*	*H*	1	−1	−1	**Max**	***2 (p = 1/2)***
	T	−1	1	−1	**Max**	***2 (1−p = 1/2)***
Column Max		−1 **Min**	−1 **Min**			
DIFFERENCE (*optimum ratio1/2*)		2 ***(q= 1/2)***	2 ***(1/q = 1/2)***			

(B) A graphical representation: saddle points do not exist

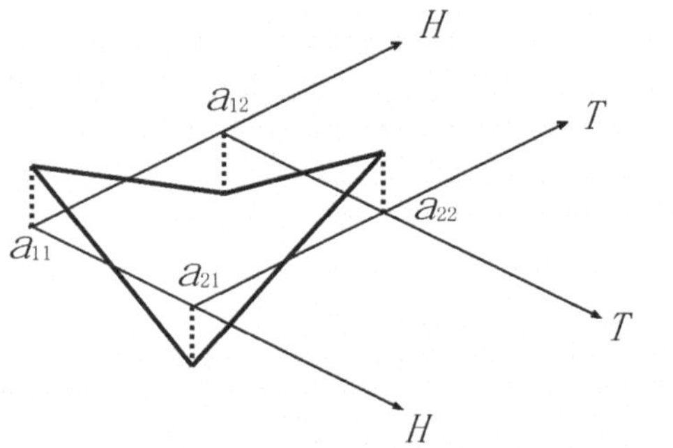

Fig. 2.4 The game of matching pennies (Case 1): the standard case

$$v_B = \text{Min}_j\ \text{Max}_i\ a_{ij} = 1. \tag{2.5}$$

Apparently, the two values, v_A and v_B, are not equal. This implies that in "The Standard Form," which limits our analysis to pure strategies, does not yield any solution at all. In other words, as Chart (B) tells us, there are no saddle points in the chart. In fact, while the two points a_{11} and a_{22} represent maximal peaks, the two points a_{12} and a_{21} indicate minimal bottoms. None of those four points can be attractive saddle points!

Now, we are caught in a sort of dilemma. Probably, there would be two different ways to get out of it. First, we would accept non-existence of the solution as such, thus deeply introspecting its philosophical meaning. Second, we would not feel self-satisfied with such non-existence, but attempting to break the deadlock by expanding the limits of solutions.

It is von Neumann and Morgenstern who have taken the second way out. More specifically, they dare to assume that each player can play a *mixed strategy* in the sense that he takes a *stochastic mixture* of the two strategies "*H*" (Head) and "*T*" (Tail). For instance, if *A* chooses *H* with probability p, and *T* with probability $(1 - p)$, *A*'s mixed strategy may be written as 【*H*, *T*: p, $1 - p$】. Likewise, if *B* chooses *H* with probability q, and *T* with probability $(1 - q)$, *B*'s mixed strategy is formulated as

(A) Stochastic mixed strategies and the payoff matrix

		Player *B*				
		q = 0	1/2	1	Row Min	
	p = 0	1	0	−1	−1	
Player *A*	1/2	0	0	0	0	Max
	1	−1	0	1	−1	
Column Max		1	0	1		
			Min			

(B) A graphical representation: b_{22} is a saddle point.

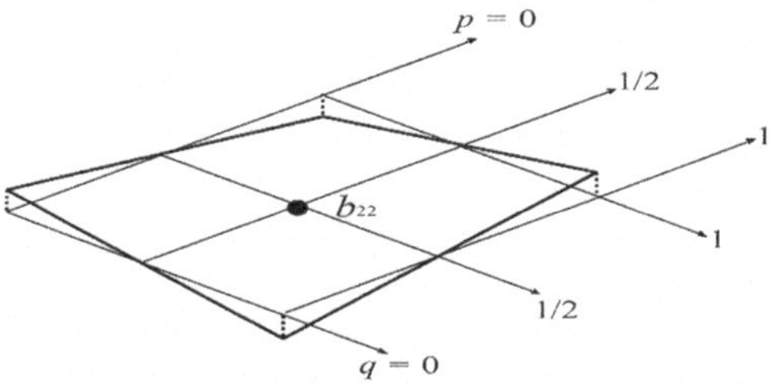

Fig. 2.5 A mixed-strategy complication of Case 1. A solution surely exists

【*H*, *T*: *q*, 1 − *q*】. If we introduce mixed strategies into the game, then the key problems to ask would be what the *optimum strategy* is all about and how to find it.

Fortunately, a very convenient method to find the optimum strategy is available. Let us take a careful look at a pair of numbers written on the right of the dotted line and at another pair on the outside of the bottom line. On the one hand, when Player *A* chooses strategy *H*, its payoff is 1 if *B* also chooses *H*, and it is (−1) if *B* chooses *T*. Therefore, we find that *the difference of the two payoffs* must be 2 in terms of the *absolute value*. On the other hand, when *A* chooses *T*, its payoff is (−1) if *B* chooses *H*, and it is 1 if *B* also chooses *T*, whence *the difference of the two outcomes* must be 2 in terms of the *absolute value*. If we compare those two values, then we immediately find that the corresponding *optimal ratios of A*'s *strategies* should be $p = 1/2$ and $1 - p = 1/2$. In an analogous fashion, we can find that the corresponding optimal ratios of *B* should be $q = 1/2$ and $1 - q = 1/2$.[5]

For convenience, let us suppose that Player *A*'s mixed strategies are limited to three: namely, $p = 0, 1/2, 1$. Similarly, Player *B*'s strategies are also limited to three: namely, $q = 0, 1/2, 1$. Then, the corresponding outcome matrix (b_{ij}) is given by Chart (A) of Fig. 2.5. If we refer to Chart (A) of Fig. 2.4 again, then we can compute the values of each b_{ij} in the following way.

[5] For details, see Suzuki (1959) and Shimizu (1997).

$$b_{ij} = (1)\,pq + (-1)\,p(1-q) + (-1)(1-p)q + (1)(1-p)(1-q). \tag{2.6}$$

For instance, if $p = 0$ and $q = 0$, then we can compute b_{11} as follows:

$$b_{11} = (1)(0)(0) + (-1)(0)(1) + (-1)(1)(0) + (1)(1)(1) = 1.$$

The remaining values of the outcome matrix in Chart (A) can be computed in an analogous way.

Now, regarding the new outcome matrix (b_{ij}), we can find the *values of the game* for Players *A* and *B* as follows:

$$v_A = \text{Max}_i \text{ Min}_j\, b_{ij} = 0, \tag{2.7}$$

$$v_B = \text{Min}_j \text{ Max}_i\, b_{ij} = 0. \tag{2.8}$$

As is quite clear, the two values, v_A and v_B, are just equal, implying that the *value of the game* is simply confirmed as follows:

$$v = v_A = v_B = 0. \tag{2.9}$$

Chart (B) tells us that the middle point b_{22} stands for a charming "saddle point" or the "equilibrium point" of the game under question. In fact, given any zero-sum two-person game, if we deal with it in an *expanded game* by introducing mixed strategies, we can always find a solution of the game in a theoretical framework, namely a saddle point in a graphical framework. This is no less than the primary purpose of writing the great book *Theory of Games and Economic Behavior* by the collaboration of mathematician Von Neumann and economist Oskar Morgenstern.

Now, let us turn to Case 2. As is seen in Fig. 2.6, this is a "biased case" in that matching on heads gives a *double premium*. Hence, the outcome matrix of Fig. 2.6 differs from the one of 2.4 by the doubling of its element a_{11}. Apparently, as both Chart (A) and Chart (B) can tell us, limiting our analysis on pure strategies yields neither a solution nor a saddle point. Although this is exactly the same result as Case 1, we can nevertheless see a difference between Cases 1 and 2. Since the element a_{11} of Case 2 is now doubled, the corresponding point in Chart (B) should tower twice higher, so that the *optimal ratios of each player's mixed strategies* should also be changed accordingly.

More specifically, following the convenient method mentioned above, we can easily compute such optimal ratios of mixed strategies. As Chart (A) of Fig. 2.6 tells us, we are able to derive $p = 2/5$ and $q = 2/5$. For illustrative convenience, let us limit the mixed strategies available to both Players *A* and *B* to the following six:

$$p \text{ or } q = 0, 1/5, 2/5, 3/5, 4/5, 1. \tag{2.10}$$

(A) Payoff matrix: limiting pure strategies does not yield solutions

		Player *B* *H*	*T*	**Row Min**		***DIFFERENCE*** *(optimum ratio)*
Player *A*	***H***	2	−1	[−1]	**Max**	***3 (p = 2/5)***
	T	−1	1	[−1]	**Max**	***2 (1 − p = 3/5)***
Column Max		[2]	[1] **Min**			
DIFFERENCE *(optimum ratio)*		3 ***(q = 2/5)***	2 ***(1/q = 3/5)***			

(B) A graphical representation: saddle points do not exist

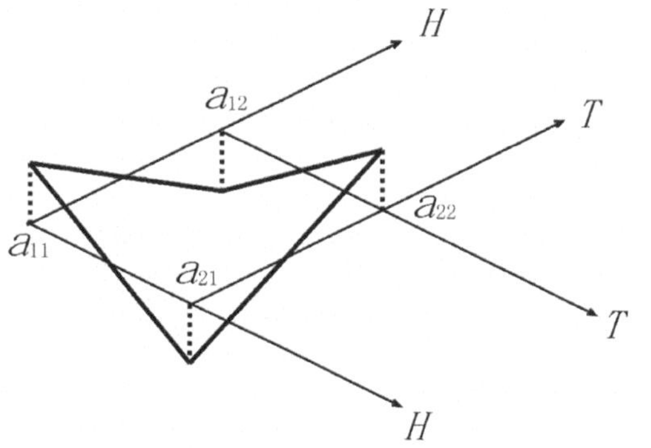

Fig. 2.6 The special game of matching pennies (Case 2). The H–H matching doubles the prize

Then, the relevant outcome matrix (b_{ij}) will be given by the 6 by 6 matrix in Fig. 2.7, in which each outcome is multiplied by 5 here. For example, if $p = 2/5$ and $q = 2/5$, then Player *A*'s outcome is given as follows:

$$b_{33} = (2)(2/5)(2/5) + (-1)(2/5)(3/5) + (-1)(3/5)(2/5) + (1)(3/5)(3/5) = 1/5.$$

Hence, multiplying the outcome by 5, we may obtain 5 $b_{22} = 1$ as we wish. Besides, we can find the *values of the game* for Players *A* and *B* as follows:

$$5v_A = 5v_B = 1.$$

Therefore, a solution surely exists in the *expanded game with mixed strategies* and is shown by $v = 1/5$. Graphically speaking, it is also a charming saddle point. Let us make a comparison between Example 1 and Example 2. This way, we will be able to well analyze the impact of giving a double premium for matching heads.

Now, we are ready to turn our attention to Example 3. As is clearly seen in Fig. 2.8, this is a "further twisted game" in which matching on heads gives a *double premium*, but failing to match on a choice (by Player *A*) of heads gives a *triple*

		Player B							
		q = 0	1/5	2/5	3/5	4/5	1	Row Min	
Player *A*	p = 0	5	3	1	−1	−3	−5	−5	
	1/5	3	2	1	0	−1	2	−2	
	2/5	1	1	1	1	1	1	1	Max
	3/5	−1	0	1	2	3	4	−1	
	4/5	−3	−1	1	3	5	7	−3	
	1	−5	−2	1	4	7	10	−1	
	Column Max	5	3	1	4	7	10		
				Min					

Fig. 2.7 Allowing mixed strategies to find a solution of the game of Case 2. For convenience, all the payoffs are quintupled here

(A) Payoff matrix: limiting pure strategies does not yield solutions

		Player *B*				
		H	*T*	Row Min		***DIFFERENCE*** *(optimum ratio)*
Player *A*	*H*	2	−3	−3		**5 (*p* = 2/7)**
	T	−1	1	−1	Max	**2 (*1 − p* = 5/7)**
	Column Max	2	1			
			Min			
	DIFFERENCE *(optimum ratio)*	3 (***q* = 4/7**)	2 (***1/q* = 3/7**)			

(B) A graphical representation: saddle points do not exist

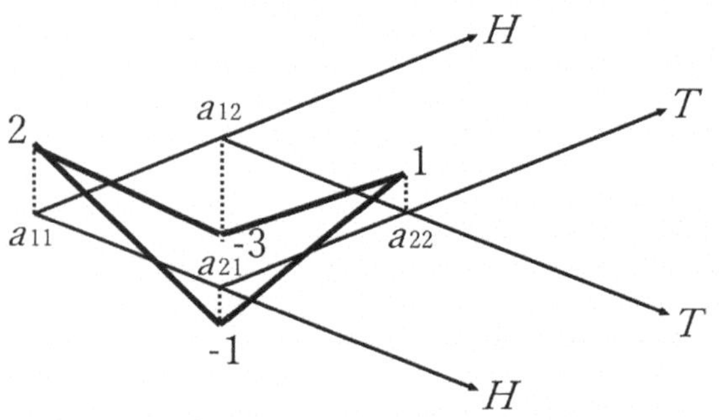

Fig. 2.8 Another special game of matching pennies (Case 3). While H–H matching doubles the prize, H–L matching triples the penalty

penalty. Interestingly enough, this game is "*a game twisted in both positive and negative ways, with a penalty overpowering a premium.*"

Let us take a look at Fig. 2.8. Then, like in previous two cases, limiting our analysis on pure strategies guarantees neither the existence of a solution nor a charming saddle point. Following the convenient method aforementioned, we can obtain the optimal ratios as follows:

$$p = 4/7, 1 - p = 3/7; q = 4/7, 1 - \mathrm{q} = 3/7.$$

For simplicity, let us limit the possible values of p and q to the following eight:

$$p \text{ or } q = 0, 1/7, 2/7, 3/7, 4/7, 6/7, 1.$$

Then, the outcome matrix of mixed strategies will be shown in Fig. 2.8, in which each outcome is multiplied by 7. For instance, if $p = 2/7$ and $q = 4/7$, then Player A's outcome is provided as follows:

$$b_{35} = (2)(2/7)(4/7) + (-3)(2/7)(3/7) + (-1)(5/7)(4/7) + (1)(5/7)(3/7) = -1/7.$$

Hence, we obtain $7b_{35} = -1$ as we wish. Other elements of the outcome matrix (b_{ij}) will be found in a similar way. Moreover, we can find the *values of the game* for Players A and B as follows:

$$7v_A = 7v_B = -1.$$

Therefore, a solution surely exists in the *expanded game with mixed strategies* and is shown by $v = -1/7$. Graphically speaking, it is also a charming saddle point. A comparison between Examples 2 and 3 will lead to a series of interesting results. To save the space, however, we omit them altogether here (Fig. 2.9).

	Player *B* q = 0	1/7	2/7	3/7	4/7	5/7	6/7	1	Row Min	
Player *A* p = 0	7	5	3	1	−1	−3	−5	−7	−7	
1/7	3	2	1	0	−1	−2	−3	−4	−4	
2/7	−1	−1	−1	−1	[−1]	−1	−1	−1	[−1]	Max
3/7	−5	−4	−3	−2	−1	0	1	2	−5	
4/7	−9	−7	−5	−3	−1	1	3	5	−9	
5/7	−13	−10	−7	−4	−1	2	5	9	−13	
6/7	−17	−11	−9	−5	−1	3	7	11	−17	
1	−21	−19	−11	−9	−1	4	9	14	−21	
Column Max	7	5	3	1	[−1] Min	4	9	14		

Fig. 2.9 Allowing mixed strategies to find a solution of the game of Case 5. For convenience, the payoffs are all multiplied by 7

2.5 Novelist Edgar Alan Poe on "The Game of Even and Odd Numbers"

As is now well known, the birth and development of game theory have been closely related to detective or mystery stories. In his 1928 paper, Morgenstern (1928), economist Oskar Morgenstern paid attention to the famous novel *The Final Problem* (1893) by writer Conan Doyle, discussing "the duel between detective Sherlock Holmes and Professor Moriarty" from a game theoretical point of view. Facing the fundamental difficulty to find a reasonable solution to the Holmes–Moriarty problem in the traditional framework of general equilibrium theory, he strongly felt the need to establish a brand new theory of games. Besides, a detailed discussion on this problem was done again in the great book *Theory of Games and Economic Behavior* (1944) by Morgenstern's collaboration with mathematician von Neumann.

We would strongly believe that *The Final Problem* should be a problem worthy of the name, so that it should be dealt with as the very final problem for two-person zero-sum games. So, before proceeding to that one, we would like to discuss here the second from the last problem, which is the "Game of Even and Odd Numbers" invented by American novelist Edgar Alan Poe (1809–1849). Poe is around 50 years older than Doyle and regarded as one of mystery story writers of world fame.[6]

"The Game of Even and Odd Numbers" is a very interesting game, which was first introduced by Poe (1844) in his mystery story *The Purloined Letter*. This game, characterizing Poe as a sharp-eyed observer of human behavior, is concerned with the game-theoretic situation in which each of the two persons is expected to outguess the strategy of his/her opponent. In his long yet nicely written paragraph, Poe remarked:

> The Prefect errs by being too deep or too shallow, for the matter in hand; and many a schoolboy is a better reasoner than he. I [Detective Dupan] knew one about eight years of age whose success at guessing in the game of "even and odd" attracted universal admiration. This game is a simple, and is played with marbles. One player holds in his hands a number of these toys, and demands of another whether that number is even or odd. If the guess is right, the guesser wins one; if wrong, he loses one. The boy to whom I allude won all the marbles of the school. Of course he had some principle of guessing; and this lay in mere observation and admeasurements of the astuteness of his opponents. For example, an arrant simpleton is his opponent, and, holding up his closed hand, asks, 'are they even or odd?' Our schoolboy replies 'odd,' and loses; but upon the second trial he wins, for he then says to himself that the simpleton had them even upon the first trial, and his amount of cunning is just sufficient to make him have them odd upon the second: I will therefore guess odd. He guesses odd, and wins. Now, with a simpleton a degree above the first, he would have reasoned thus: This

[6]Edgar Alan Poe was known as a bit strange person, but his passion for literature was so great that he wrote many masterpieces such as *Black Cat*, *The Fall of the House of Usher*, *The Murders in the Rue Morgue*, and *The Raven*. We would like to point it out that his influence on Japanese world of letters was very conspicuous. Even a Japanese noted novelist changed his pen name to "Edogawa Rampo," somewhat sounding like "Edgar Alan Poe." Poe's outstanding contribution to the relation between literature and mathematics has attracted a number of mathematicians including Yoshioka (1966).

> fellow finds that in the first instance I guessed odd, and, in the second he will propose to himself upon the first impulse, a simple variation from even to odd, as did the first simpleton, but then a second thought will suggest that this is too simple a variation, and finally he will decide upon putting it even as before. I will therefore guess even, and win. Now this mode of reasoning in the schoolboy, whom his fellows termed "lucky." In its last analysis, what is it? 'It is merely' I said, 'an identification of reasoner's intellect with that of his opponent.' (Poe 1844)

Remarkably, this story by Poe describes a game-theoretic situation where "a wonder boy" and his opponent, either "an arrant simpleton" or "a bit clever boy," play the simple "Game of Even and Odd Numbers." Each player has only two strategies, "Even (Number)" and "Odd (Number)." As is shown in Fig. 2.10, there are typically two trials of games. At the first trial, the opponent Simpleton holds in his hand a number of marbles, asks Wonder Boy whether the number is "Even" or "Odd." Suppose that Wonder Boy replies "Odd" and loses, with his outcome being (−1). Then, the critical second trial will begin. Presumably, there are two different kinds of opponents, an Arrant Simpleton and a Clever Kid. On the one hand, if the opponent is an Arrant Simpleton, he is expected to simply change his strategy from "Even" to "Odd," and Wonder Boy rightly guesses "Odd" and wins. On the other

【I. The First Trial】

THE WONDER BOY		THE OPPONENT: [EVEN]	ODD
	EVEN	1	−1
	[ODD]	[−1]	1

The Wonder Boy loses and the opponent wins

…

【II (A). The Second Trial against a mere simpleton】

If THE OPPONENT is *an arrant simpleton,* then THE WONDER BOY simply changes his strategy:
EVEN ⇒ ODD

THE WONDER BOY		THE OPPONENT: [EVEN]	⇒ ODD
	EVEN	1	−1
	[ODD]	−1	[1]

The Wonder Boy wins

【II (B). The Second Trial against for a little cleverer kid】

If the opponent is *a little cleverer kid*, then the wonder boy has to take a more complicated strategy. Although the boy might be tempted to change his strategy from **EVEN to ODD,** he must have a second thought to change his strategy again from **ODD** to **EVEN, namely EVEN ⇒ODD ⇒EVEN.**

THE OPPONENT: EVEN ⇒ ODD; [EVEN] ⇐ ODD as a second thought

THE WONDER BOY		EVEN	ODD
	[EVEN]	[1]	−1
	ODD	−1	1

Also The Wonder Boy wins

Fig. 2.10 Edgar Alan Poe on the Game of Even and Odd Numbers

hand, if the opponent is a Clever Kid, his next strategy becomes more complicated than an Arrant Simpleton. The Clever Kid will have the *second thought* that a simple variation from "Even" to "Odd," as did the first Simpleton, is too simple a variation to believe, and finally, he will change again to choose "Even" again. As a result, the Wonder Boy's guessing is "Old" and eventually wins. Consequently, whether the opponent is Simpleton or Clever Kid, the Wonder Boy eventually wins the game.

Undoubtedly, Poe's Game of Even and Odd Numbers is no less than a Game of Matching Pennies, its outcome matrix being identical to that of Fig. 2.4. It is in this sense that Poe may be regarded as one of pioneers of Game Theory. In Poe's Game, as was repeatedly discussed above, limiting our analysis on pure strategies guarantees neither a solution nor a saddle point. If we expand the scope of strategies by allowing mixed strategies, however, as was shown by Fig. 2.5, a solution (or a saddle point) surely exists when each player chooses a specific mixed strategy of "Even" and "Odd" with a 50–50 chance. Then, the solution of the game would be just zero. Hence, if this kind of game is played repeatedly, the game would be fair in the sense that each player would be a winner or a loser with a 50–50 chance. Hence, there should be neither a one-sided winner nor a one-sided loser.

Here, we find that Poe's sharp observation is in sharp contrast to the Neumann–Morgenstern way of thought. According to Poe, if the Game of Even and Odd Numbers is played repeatedly, then the winning ratio (or the losing ratio) of each player will never converge to a half, but tends either to expand to unity or to contract to zero. As a result, as is seen in Fig. 2.10, the "wonder boy" will dominate the game and monopolize all the prizes, whereas the "less intelligent kid" will be exploited by the "wonder boy." Such an unfair distribution of the prizes, according to Poe as a critical observer, originates in unequal distribution of intelligence and anticipation between the two players.

In short, Edgar Alan Poe, as a critical observer of the real world, has constantly emphasized the cruel fact that in any two-person game, the coexistence of a one-sided winner and a one-sided loser appears. This radically differs the situation game theorists have ardently advocated. Therefore, we might see a dilemma between "the practical solution" and "the theoretical solution." We wonder which solution the reader would be inclined to support. In our opinion, Poe's objection against an easy theoretical solution is really worth paying attention to. In general, theory is no more than theory and should not be almighty. If there occurs an irreconcilable discrepancy between the theory and the reality, we must rely on the reality, not on the theory.

2.6 Conan Doyle's Story "The Final Problem" as a Variant of Matching Pennies

We are now in a position to reexamine Conan Doyle's story *The Final Problem* as a variant of matching pennies. Although we already discussed it in Chap. 1 of this book, we wish to shed new light on it from the viewpoint of the Game of Matching

(A) Payoff matrix: limiting pure strategies does not yield solutions

		HOLMES Dover	HOLMES Canterbury	Row Min	
MORIARTY	Dover	2	0	[0]	Max
	Canterbury	−1	2	−1	
	Column Max	[2] Min	[2] Min		

(B) Allowing mixed strategies to find a solution of the game (A). For convenience, all the payoffs are multiplied by five.

		HOLMES $q = 0$	1/5	2/5	3/5	4/5	1	Row Min	
MORIARTY	$p = 0$	10	7	4	1	−2	−5	−5	
	1/5	8	6	4	2	0	−2	−2	
	2/5	6	5	4	3	2	1	1	
	3/5	4	4	[4]	4	4	4	[4]	Max
	4/5	2	3	4	5	6	7	2	
	1	0	2	4	6	8	10	0	
	Column Max	10	7	[4] Min	6	8	10		

Fig. 2.11 The Final Problem as a Game of Matching Pennies (FIRST Interpretation): Showdown between Moriarty and Holmes

Pennies. Exactly speaking, in von Neumann and Morgenstern (1944), *The Final Problem* is regarded as Example 4, just following the previous Examples 1, 2, and 3.

In *The Final Problem*, there are the two players, Professor Moriarty and Sherlock Holmes. While their levels of intelligence are almost equal, their strategies should be different, thus requiring cool manipulation and shrewd anticipation. As is seen in Fig. 2.11, each person has two strategies, "Proceeding to Dover" and "Quitting at Canterbury. The fields (D,D) and (C,C) correspond to Moriarty catching Holmes, which is reasonable to describe by a certain positive value. Whereas the field (C,D) shows that Holmes successfully escaped to Dover (and presumably to the Continent), Moriarty had to stop at the intermediate station Canterbury. Clearly, this means Moriarty's defeat, thus being describable by a certain negative value. Finally, the field (D,C) means that Holmes escapes Moriarty at Canterbury, but fails to reach the Continent. This can be thought of as a draw.

The most important problem to ask here is specifically what numbers should be filled in the four fields, namely (D,D), (C,C), (D,C), and (C,D). Conceivably, there would be a variety of ways of giving numbers. The FIRST interpretation by von Neumann and Morgenstern (1944) is given by Fig. 2.11.

As is plainly clear, the outcome matrix in Chart (A) of Fig. 2.11 shows that overall, Professor Moriarty has an advantage over Sherlock Holmes in their confrontation. In fact, whereas the positive value of (+2) appears twice, namely in the fields (D,D) and

(C,C), the negative value of (−2) does exist at all, and the negative value (−1) appears only once in the field (C,D), with the zero value in the field (D,C).[7]

As can easily be seen, limiting our analysis on pure strategies does not guarantee a solution of the game. This is because the following inequality has to hold here:

$$v_M = \mathrm{Min}_i\ \mathrm{Max}_j\ a_{ij} = 2 > 0 = \mathrm{Max}_j\ \mathrm{Min}_i\ a_{ij} = v_H. \tag{2.11}$$

The value of the game for Moriarty is greater than that for Holmes, implying that the game per se is not fair at all in the sense that it is actually more advantageous to Moriarty than Holmes.

Now, in line with the idea of von Neumann and Oskar Morgenstern, let us expand the scope of strategies of each player to introduce the "mixture of pure strategies," or simply "mixed strategies" into the game. Specifically, let us consider the mixed strategy of mixing the pure strategies D and C, with the ratio p and $(1 - p)$, which can be denoted by $[D, C; p, 1 - p]$. For the sake of convenience, let us limit the values of p and q on the five numbers: 0. 1/5, 2/5, 3/5, 4/5, and 1. Then, we obtain the outcome matrix of mixed strategies (b_{ij}), as is seen in Fig. 2.11.

In this expanded game, the pair of mixed strategies $(p, q) = (3/5, 2/5)$ gives a solution, thus being a saddle point. For convenience, each value of Panel (B) is multiplied by 1/5. Then, it is an easy job to derive the following equation:

$$v_M = \mathrm{Min}_i\ \mathrm{Max}_j\ b_{ij}/5 = 4/5 = \mathrm{Max}_j\ \mathrm{Min}_i\ b_{ij}/5 = v_H. \tag{2.12}$$

Since the game of the expanded game is a positive value, we are confirmed again that the game per se is an unfair game, guaranteeing a one-sidedly victory for Moriarty and a one-sided defeat for Holmes. Moriarty's best strategy is the strategy of proceeding to Dover with probability 60%, whereas Holmes' best strategy is the same strategy with probability 40%. In plain English, while Moriarty has to make a special lottery containing "three Dover balls and two Canterbury balls", Holmes must make another symmetric lottery containing "two Dover balls and three Canterbury balls." Of course, those lotteries should be free and fair, being independent of any possible interferences. Although such lot drawing is perhaps conceivable in a fiction, it never exists in the real world in which flesh-and-blood persons like ourselves live. It is in this sense that the validity of introducing mixed strategies into the game must be discussed very seriously.

Now, as the *second* possible interpretation, let us consider Fig. 2.12 in which, contrary to the *first* interpretation, Sherlock Holmes is in a more advantageous position than Moriarty. If we decease *every outcome* in Panel (A) of the last Fig. 2.11 by the equal amount of one unit, then we immediately obtain the outcome matrix in Panel (A) of the present Fig. 2.12. Then, the positive value (+2) is now

[7] In von Neumann and Morgenstern (1944), big numbers such as "plus 100" and "minus 50" were conspicuously employed. For analytical convenience, however, those numbers are multiplied by the fraction (1/50) here.

(A) Limiting pure strategies does not yield solutions. For convenience, all the payoffs are multiplied by (1/50) here.

		HOLMES Dover	HOLMES Canterbury	Row Min	
MORIARTY	Dover	1	−1	−1 (boxed)	Max
	Canterbury	−2	1	−2	
	Column Max	1 (boxed) Min	1 (boxed) Min		

(B) Allowing mixed strategies to find a solution of the game (A) Remark. For convenience, all the payoffs are multiplied by five.

		HOLMES $q = 0$	1/5	2/5	3/5	4/5	1	Row Min	
MORIARTY	$p = 0$	5	2	−1	−4	−7	−10	−10	
	1/5	3	1	−1	−3	−5	−7	−7	
	2/5	1	0	−1	−2	−3	−4	−4	
	3/5	−1	−1	−1 (boxed)	−1	−1	−1	−1 (boxed)	Max
	4/5	−3	−2	−1	0	1	2	−3	
	1	−5	−3	−1	1	3	5	−5	
	Column Max	5	2	−1 (boxed) Min	1	3	5		

Fig. 2.12 The Final Case as a Game of Matching Pennies (Second Interpretation)

replaced by a smaller value (+1) in the fields (D,D) and (C,C), and the negative value (−) is lessened to (−2). Not only that, the positive value (+2) is also reduced to (+1). Overall, Holmes' position is considerably improved in comparison with the last situation.

As is quite clear in Panel (A) of Fig. 2.12, limiting on pure strategies does not guarantee a solution of the game. If we introduce mix strategies into the matrix, it is clear in Panel (B) that a solution surely exists, with $(p, q) = (3/5, 2/5)$ becoming a saddle point as before. The solution, (−1/5) of the expanded game, certainly confirms Holmes' advantage against Moriarty.

Finally, let us attempt to give the *third* interpretation, by which the Game of *The Final Problem* is completely fair for either player. Concerning the two players, one player has neither an advantageous nor a disadvantage over the other.

Let us take a look at Panel (A) of Fig. 2.13. Then, we can see that there are the two positive values, (+2) and (+1), and the two negative values, (−2) and (−1), showing the fair distribution of outcomes in the matrix. In this game, limiting our analysis on pure strategies does not guarantee a solution at all. When we are allowed to take care of mixed strategies in the expanded game, however, a solution surely exists and is indicated by a saddle point $(p, \text{q}) = (2/3, 2/3)$. Note that the value of the game is

(A) Limiting pure strategies does not yield solutions

		HOLMES			
		Dover	**Canterbury**	**Row Min**	
MORIARTY	**Dover**	1	−1	**−1**	**Max**
	Canterbury	−2	2	**−2**	
	Column Max	**1**	**2**		
		Min	**Min**		

(B) Allowing mixed strategies to find a solution of the game (A). For convenience, all the payoffs are multiplied by six.

		HOLMES								
		$q = 0$	**1/6**	**2/6**	**3/6**	**4/6**	**5/6**	**1**	**Row Min**	
MORIARTY	$p = 0$	12	8	4	0	−4	−8	−12	**−12**	
	1/6	9	6	3	0	−3	−6	−9	**−9**	
	2/6	6	4	2	0	−2	−4	−6	**−6**	
	3/6	2	2	1	0	−1	−2	−3	**−3**	
	4/6	0	0	0	0	0	0	0	**0**	**Max**
	5/6	−3	−2	−1	0	1	2	3	**−3**	
	1	−6	−4	−2	0	2	4	6	**−6**	
	Column Max	**12**	**8**	**4**	**0**	**2**	**4**	**6**		
					Min					

Fig. 2.13 The Final Problem as a Game of Matching Pennies (Third Interpretation)

$v = 0$. This clearly demonstrates that this game is a completely fair game for both players, namely Moriarty and Holmes.

In the above, we have given three different interpretations of *The Final Problem* as a variant of "The Game of Matching Pennies." They are the First, Second, and Third interpretations. Unfortunately, von Neumann and Morgenstern (1944) discussed only the First interpretation, thus completely neglecting the Second and Third interpretations. We do think, however, that such unbalanced treatment is fairly groundless and should be amended. In our opinion, the Second and Third interpretations should be as important as the First interpretation.

We are entitled to choose any one interpretation. Whichever interpretation we may choose, there nevertheless exists a hardly surmountable barrier before us. This barrier is related to the practical difficulty to determine the best mixed strategy from the relevant outcome matrix. On the one hand, knowing that Holmes is coming soon, Moriarty has to make a quick decision by making a strange lottery box including "three Dover balls and two Canterbury balls." We do not think that this is practically feasible for Moriarty as a "gentleman of high pride." Presumably, his pride would be seriously wounded and hardly recoverable. Similarly for Sherlock Holmes as an archrival of Moriarty. They are engaging in an only once game that cannot be repeated. In such a life-or-death situation, relying on lotteries would hardly be conceivable. On the other hand, the situation in which the two persons of equal intelligence, like Moriarty and Holmes, are playing a life-or-death game would be exceptionally rare. In almost all situations, as is seen in the confrontation between highly intelligent Sherlock Holmes and arrant simpleton Cornel Moran, the IQs of the two opponents are likely to be quite different. Then, the result of the game would be what Poe's Game of Even and Odd Player would eloquently tell us: that is, one player will be a dominant winner and the other player a miserable underdog.

To sum up, when we straightforwardly apply game theories to real situations, we should distinguish between the two different kinds of games. One kind of games, like outdoor games including baseball and tennis, are repeatable and regarded as "light games." The other games are more serious and non-repeatable, thus being thought of as "heavy games." It is in those "heavy games" that the validity of adopting mixed strategies would be seriously challenged. Further investigations on this important problem would be required.

2.7 Reassessment of "The Game of Stone, Paper, Scissors" as Another Variant of Matching Pennies

We are now in a position to discuss another zero-sum two-person game, which has been favored by von Neumann and Morgenstern. It is nostalgically called "The Game of Stone, Paper, Scissors." It is conventional for us to think that "Stone defeats Scissors," "Scissors defeat Paper," and "Paper defeats Stone." Needless to say, there should be no ethical judgments on "Stone," "Paper," and "Scissors" per se.

It is quite easy to understand that "Matching Pennies" is regarded by von Neumann and Morgenstern as a very important illustration of zero-sum games, since it has been very popular in everyday life in any Western society. "Stone, Paper, Scissors," however, seems to be not so popular as "Matching Pennies." There might be some people asking why those two games require equal treatment. In this respect, von Neumann and Morgenstern (1944) once remarked:

> These two examples show the difficulties which we encounter in a not-strictly determined games, in a particularly clear form; just because of their extreme simplicity the difficulty is perfectly isolated here, in vitro. The point is that in "Matching Pennies" and in "Stone, Paper, Scissors," any way of playing is just as good as any other: There is no intrinsic advantage or disadvantage in "heads" or "tails" per se, nor in "stone," "paper" or "scissors" per se. The only thing which matters is to guess correctly what the adversary is going to do; but how are we going to describe that without further assumptions about the players' "intellect"? There are, of course, more complicated games which are not strictly determined and which are important from various more subtle, technical points. But as far as the main difficulty is concerned, the simple games of "Matching Pennies" and of "Stone, Paper, Scissors" are perfectly characteristic. (von Neumann and Morgenstern 1944, p. 111)

In historical perspective, at Princeton University in the 1940s, Dr. Shizuo Kakutani, then a Japanese rising star, worked hard as a research assistant to Professor John von Neumann. This historical fact gives the conjecture that Kakutani personally told von Neumann and Morgenstern "the Exciting Game of Stone, Paper,

(A) Limiting pure strategies does not yield solutions

		Player *B*				
		Stone	Paper	Scissors	Row Min	
Player *A*	Stone	0	−1	1	−1	Max
	Paper	1	0	−1	−1	Max
	Scissors	−1	1	0	−1	Max
	Column Max	1	1	1		
		Min	Min	Min		

(B) A graphical representation: saddle points do not exist

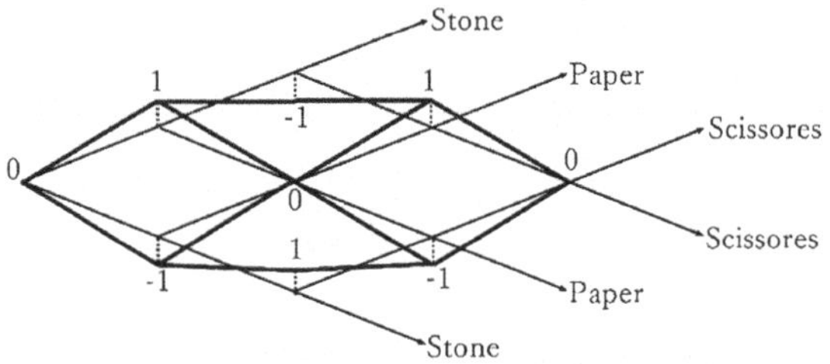

Fig. 2.14 The "Final Problem" as a Game of Matching Pennies (Third Interpretation): Limiting pure strategies does not yield "saddle point solutions"

Scissors." This conjecture might be correct or not correct, requiring still further investigation.[8]

The outcome matrix of pure strategies for "The Game of Stone, Paper, Scissors" can be given in Fig. 2.14. Each of the two players, Players *A* and *B*, has the three strategies; "Stone," "Paper," and "Scissors." If we denote "win" by "+1," "loss" by "−1," and "draw" by "0," the relevant outcome matrix is shown by Fig. 2.14. As is easily seen in Upper Chart (A), the following inequality holds:

$$v_A = \mathrm{Min}_i\ \mathrm{Max}_j\ a_{ij} = 1 > -1 = \mathrm{Max}_j\ \mathrm{Min}_i\ a_{ij} = v_B. \tag{2.13}$$

Hence, limiting our analysis on pure strategies guarantees neither a solution nor a saddle point. Lower Chart (B) graphically shows the non-existence of a saddle point. Now, let us expand the scope of strategies to allow the mixture of pure strategies, namely *mixed strategies*. Suppose that Player *A* chooses the mixed strategy "Stone with probability p_1, Paper with probability p_2, Scissors with probability p_3," where

[8]Later, Kakutani became a famous mathematician who succeeded in establishing a generalization of Brouwer's fixed point theorem. This generalization theorem was effectively used to prove the existence of general equilibrium. For this point, see Neumann and Morgenstern (1944), page 154, long footnote 1. Personally speaking, a brief meeting with Kakutani at Rochester in the 1960s has been in my fond memory until today.

		Player *B*				
		(100)	(1/3)	(010)	(001)	Row Min
Player *A*	(100)	0	0	−1	1	−1
	(1/3)	0	[0]	0	0	[0] Max
	(010)	1	0	0	−1	−1
	(001)	−1	0	1	0	−1
Column Max		1	[0] Min	1	1	

Fig. 2.15 Allowing mixed strategies in the Game of Stone, Paper, Scissors surely yields a solution

$p_1 + p_2 + p_3 = 1$, and that Player *B* chooses the mixed strategy "Stone with probability q_1, Paper with probability q_2, Scissors with probability q_3," where $q_1 + q_2 + q_3 = 1$. Then, the outcome of the mixed strategy of Player *A* is written as follows:[9]

$$b_{ij} = \sum_i \sum_j p_i \, a_{ij} \, q_j. \tag{2.14}$$

Given a_{ij}'s, the value of b_{ij} depends on Player *A*'s strategy vector $p = (p_1, p_2, p_3)$, and Player *B*'s strategy vector $q = (q_1, q_2, q_3)$. As can naturally be expected, the optimal mixed strategies of the two persons are provided as follows:

$$p = (1/3, 1/3, 1/3) \text{ and } q = (1/3, 1/3, 1/3). \tag{2.15}$$

In other words, Player *A*'s optimal strategy must be the *fair strategy* in the sense that he chooses "Stone," "Paper," and "Scissors" with *equal probability*, that is, 1/3. In order to make this point clearer, let us simply write (1/3,1/3,1/3) as (1/3), (1,0,0) as (100), (0,1,0) as (010), and (0,0,1) as (001). Then, if we dare to ignore all other mixed strategies except those four, then the matrix of mixed strategies for "The Game of Stone, Paper, Scissors" will be given in Fig. 2.15. It is an easy job to see in Panel (A) that a solution of the mixed game exists and given by $v = v_A = v_B = 0$. Besides, we can show that the solution mentioned above is an attractive saddle point. To save the space, however, its graphical representation in the three-dimensional Euclidean space will be omitted here.

To be fair, Theory is no more than Theory, and nothing else. Sometimes (even too often), however, Apparent Reality may differ from True Reality. Theoretically speaking, the Game of Stone, Paper, Scissors should be a *fair game* in the sense that it is neither advantageous nor disadvantageous to either player. If we play that game repeatedly, the winning probability (and also the losing probability) will tend to a half.

[9] Since this is a zero-sum game, it follows that Player *B*'s outcome is given as follows: $b_{ij} = \Sigma_i \Sigma_j p_i \, (1 - a_{ij}) q_j$.

Fortunately or unfortunately, however, such fairness is not likely to be supportable in most of daily games. In the real world, we often eyewitness the eventual dominance of a "wonder boy," who will tend to defeat many other "simpletons" after several rounds. The question of much interest is then why in the real world, a one-sided winning streak for the wonder boy and a one-sided losing streak for a simpleton are likely to happen.

This may sound strange to some people. We do think, however, that a simpleton who is overanxious to win is doomed to defeat, whereas a wonder boy who is highly intelligent and very good at outwitting is a perfect winner. It would be an ironic consequence to see that a robot-like simpleton with no ambition for victory would do a better job in the simple Game of Stone, Paper, Scissors. Believe or not, this is likely to agree with common sense.

2.8 Common Sense and Mathematical Rigor: Good Advice from von Neumann and Morgenstern

In the long history of Japanese Art and Sciences, the following maxim has been handed down the generations:

> Never forget the ideals with which you started out. (Ze-a-mi 1440)

This maxim was first told by Ze-a-mi (1363–1443), a famous Noh drama writer in pre-modern Japan. Today, it seems that the specter of Game Theory is haunting the economics profession and haunting indeed with no relation to the original ideals of its founders, i.e., John von Neumann and Oskar Morgenstern. It is high time for us to get back to the original ideas, which we started out. Otherwise, we would aimlessly be going up in higher and higher in the wind like a kite without its string.[10]

As was mentioned above, there are two memorable years for Game Theory. The first important year was 1928 as the Year of Birth, in which Neumann's paper on social game theory and Morgenstern's paper on economic forecasting were independently published. While the so-called Min-Max Theorem for zero-sum games was first established and proved by Neumann, the Holmes–Moriarty duel in Conan Doyle's Sherlock Holmes stories was first noticed as a "troublesome game" by Morgenstern. The second important year for Game Theory was 1944 as the Year of Maturity, in which the great book *Theory of Games and Economic Behavior* was finally published. This book was a bulky book with more than 600 pages, has been a sort of "Bible of Game Theory." To be honest, we have been "faithful readers of the Game Theory Classic." We now realize that there exist, and are still widening, the gaps between the "Game Classic 1944" and the apparently brilliant developments in more recent times. Thus, it is high time to get back to the original ideas, which we

---

[10]For this point, see Morishima (1994).

started out. For related topics, see Morgenstern (1958, 1972, 1976), Nakayama (1997), Nakayama et al. (2000), Nasar (1998), and Soros (2008).

In the following, we will summarize what we have learned in Chap. 1 and this chapter.

1. The very core of the Classical Game Theory is made up of "Zero-Sum Game," but not by "Non-Zero-Sum Game." As a matter of fact, it is an easy job for us to introduce an "extra person" so that we can expand a given non-zero-sum n person game into a zero-sum $(n + 1)$ person game.
2. Concerning the family of zero-sum games, the zero-sum Two-Person Games constitute the most important member.
3. Von Neumann and Morgenstern believe that the simple Games of Matching Pennies, and of Stone, Paper, Scissors are perfectly characteristic of all the two-person zero-sum games.
4. "The Final Case," featuring the Moriarty–Holmes confrontation, can be regarded as a variant of the game of matching pennies.
5. The solution of the original game with *pure* strategies and that of the expanded game with *mixed* strategies should not be discussed on the same level of abstraction. A special caution for the interpretation of the mixed strategies must be taken.
6. In order for the solution of the expanded game to be meaningful and applicable to the economic reality, the game in question must be repeatable at many times. Moreover, the two players are required to have the equal degree of intelligence and outguessing ability. In reality, those conditions are usually not satisfied.
7. Overall, the gulf between Theory and Reality is too large to fill in and gets larger and larger in such a complex world as we live in.

Finally, we must keep in mind that von Neumann and Morgenstern have left us the following warning:

> This is another instance of the characteristic relationship between common sense and mathematical rigor. . . . if the mathematical proof fails to establish the commonsense result, then there is a strong case for rejecting the theory altogether. Thus the primate of mathematical procedure extends only to establish checks on the theories—in a way which would not be open to common sense alone. (von Neumann and Morgenstern 1944, p. 361)

As our everyday experience can tell us, we often see the conflict between common sense and mathematical rigor. We are quite confident that we should rely on common sense rather than mathematical rigor. As the saying goes, simple is beautiful!

References

Aumann RJ, Hart S (eds) (1992) Handbook of game theory and economic applications. North Holland

Dasgupta P, Gale D, Hart O, Maskin E (eds) (1992) Economic analysis of markets and games. The MIT Press, Cambridge, MA

Dixit AK, Nalebuff BJ (1991) Thinking strategically: the competitive edge in business, politics and economic life. Norton, New York
Luce RD, Raiffa H (1957, rev. 1989) Games and decisions: introduction and critical survey. Wiley, New York
Morgenstern O (1928) Wirtschaftsprognose, einer Untersuchung ihrer Voraussetzungen und Mölichkeiten (English: Economic forecasting: a study of its difficulties and impossibilities). Springer, Vienna
Morgenstern O (1935) Vollkommene Voraussicht und wirtschaftliche Gleichgewicht (English: Perfect foresight and economic equilibrium). Z National 6(3):337–357
Morgenstern O (1958) John von Neumann, 1883–1957. Econ J 68:499–503
Morgenstern O (1972) Thirteen critical points in contemporary economic theory: an interpretation. J Econ Lit 10(4):1163–1189
Morgenstern O (1976) The collaboration between Oskar Morgenstern and John von Neumann. J Econ Lit 14(3):805–816
Morishima M (1994) Modern economics as economic thought. Iwanami Publishers, Tokyo
Morishima M (ed) (1998) Joseph A. Schumpeter & Yasuma Takata: power or pure economics? Macmillan Press, London
Nakayama C (1997) The process of collaboration between von Neumann and Morgenstern. History Econ Rev 26:40–50
Nakayama M, Muto S, Funaki Y (eds) (2000) Thinking in game theory (in Japanese). Keiso Publishers, Tokyo
Nasar S (1998) A beautiful mind: the life of mathematical genius and Nobel laureate John Nash. Touchstone, New York
Okada A (1996) Game theory (in Japanese), 2nd edn, 2011. Yuhikaku Publishers, Tokyo
Poe EA (1844) The purloined letter. In: The gift of a Christmas present for 1845. Cary and Hart, Philadelphia
Poundstone W (1992) Prisoner's dilemma: John von Neumann, game theory, and the puzzle of the bomb. Doubleday, New York
Sakai Y (1982) The economics of uncertainty (in Japanese). Yuhikaku Publishers, Tokyo
Sakai Y (1996) The economics of risk: information and culture (in Japanese). Yuhikaku Publishers, Tokyo
Sakai Y (2000) The concept of risk in economics: the impact of sympathy and animal spirits (in Japanese). In: Economy, society, and politics. Economic Planning Agency Publishers, Tokyo, pp 5–20
Sakai Y (2001) Von Neumann, Morgenstern, and the world of game theory: part 1 (in Japanese). Tsukuba Econ Stud, Spring: 1–42
Sakai Y (2005) On risk society (in Japanese). Iwanami Publishers, Tokyo
Sakai Y (2010) The economic thought of risk and uncertainty (in Japanese). Minerva Publishers, Kyoto
Sakai Y (2019) J.M. Keynes versus F.H. Knight: risk, probability, and uncertainty. Springer Nature, Singapore
Schotter A (ed) (1976) Selected economic writings of Oskar Morgenstern. New York University Press, New York
Shimizu Y (ed) (1997) On the theory of games: strategies and decision trees (in Japanese). Nihon Jitsugyo Publishers, Tokyo
Shubik M (1967) Essays in mathematical economics: in honor of Oskar Morgenstern. Princeton University Press
Shubik M (1982) Game theory in the social sciences: concepts and solutions. The MIT Press, Cambridge, MA
Soros G (2008) The crash of 2008 and what it means. Public Affairs, New York
Suzuki M (1959) The theory of games (in Japanese). Keiso Publishers, Tokyo
Suzuki M (1994) The new theory of games (in Japanese). Keiso Publishers, Tokyo
Suzuki M (1999) The world of game theory (in Japanese). Keiso Publihers, Tokyo
Takata Y (1937) The theory of interest (in Japanese). Iwanami Publishers, Tokyo
Takata Y (1940a) New studies in interest theory. Iwanami Publishers, Tokyo

Takata Y (1940b) Studies in race and economy (in Japanese). Yuhikaku Publishers, Tokyo
Takata T (1941) Essays in power theory (in Japanese). Iwanami Publishers, Tokyo
Takata Y (1942) On economy and race (in Japanese). Iwanami Publishers, Tokyo
Takata Y (1943) Austere living (in Japanese). Kocho Shorin Publishers, Kyoto
Takata Y (1944) On economic control (in Japanese). Tokyo, Nihon Hyoron Publishers
Takata Y (1955) Critical reassessment of John M. Keynes: from the viewpoint of power theory (in Japanese). Yuhikaku Publishers, Tokyo
Taleb NN (2007) The black swan: the impact of the highly improbable. Random House, New York
Varoufakis Y (ed) (2001) Game theory: critical concepts in the social sciences. Routledge, London
von Neumann J, Morgenstern O (1944) Theory of games and economic behavior. 2nd edn, 1946; 3rd edn, 1953. Princeton University Press
Yoshioka S (1966) Poe's short stories and mathematics (in Japanese). In: Iyanaga S et al (eds) Mathematics and human life. Gakuseisha Publishers, Tokyo
Ze-a-mi (1440) *Hu-shi-ka-den*, or the essence of Noh drama (in Japanese)

Chapter 3
Non-zero-sum Games and Nash Equilibriums: Applications to Generation Gaps Problems

Abstract This chapter critically discusses non-zero-sum games with special reference to Nash equilibrium, which is a successor of Cournot equilibrium in the theory of duopoly and oligopoly. We first focus on the "residence game" as a typical non-zero-sum game. We pick up the old and the young couples, who have to decide whether they live together or separately and whether they live in the country or in the city. We shed new light on "Generation Gaps Problems," with many other applications. We next turn to the "Battle of the Sexes," or more plainly the "dilemma of lovers." The male and the female have to find good dating spots. Whereas the male prefers to see boxing rather than ballet, the female's preference is just the opposite. We propose that contrary to the conventional way of discussions, the introduction of a *third option* such as going to see movies gives each couple a *second best solution*. Finally, comparison between "Econs" and "Humans" is carefully examined.

Keywords Nash as a successor of Cournot · The residence game as a non-zero-sum game · Generation gaps problem · Dilemma of lovers · Dilemma of prisoners · Arms race · Econs versus humans

3.1 From Antoine Cournot to John F. Nash

The main purpose of this chapter is to critically evaluate The Theory of Non-Zero-Sum Games as opposed to Zero-Sum Games and Applications, which have been extensively discussed in the last two chapters. We will start our discussion with the unique transition of equilibrium concepts from Antoine A. Cournot to John F. Nash Jr.

This chapter is a completely revised version of Sakai (1995). I am grateful to Oscar Morgenstern (Princeton), Jack Ocks (Pittsburgh), Asatoshi Maeshiro (Pittsburgh), Yoshio Kimura (Nagoya Municipal), and Mitsuo Suzuki (Tokyo Kogyo) for helpful comments and suggestions.

Y. Sakai, *Games, Decisions, and Markets*, New Frontiers in Regional Science: Asian Perspectives 74, https://doi.org/10.1007/978-981-99-5285-4_3

3.1.1 Joseph A. Schumpeter's Praise for Antoine Cournot

Joseph A. Schumpeter (1883–1950), a "giant of the twentieth century," is known as the economist who very highly rated the achievement of Antoine A. Cournot (1801–1877) in the history of economic analysis. In what follows, let us introduce the reader to a couple of episodes to clearly show how highly Schumpeter evaluated Cournot.

When Schumpeter taught economic science at Harvard University, he enthusiastically told his students about the history of economic giants. Among those attentive students were Paul A. Samuelson (1915–2001), then a young ambitious graduate student and later a Nobel Economic Science winner. Schumpeter's talk proceeded to concretely name "the Four Greatest Economists in History" and bravely gave great shocks to all the students by saying that among those four, three economists are prideful French scholars. According to Schumpeter, the following four were honorably selected. They are François Quesnay (1694–1774), Antoine A. Cournot (1801–1877), Léon Walras (1834–1910), and the fourth economist unnamed.[1]

Quesnay was a French physician and economist, being well known for publishing his work *Tableau économique* (Economic Table) in 1758. This was presumably the first work to describe the working of the national economy in an analytical fashion. Cournot was a noted French mathematician, also greatly contributing to the development of oligopoly theory as an important field of economic theory. In his 1838 book *Researches into the Mathematical Principles of the Theory of Wealth*, Cournot used differential and integral calculus, apparently a highly advanced mathematical tool at that time, causing harsh criticism in economics profession. Those criticisms against Cournot, however, promptly vanished 100 years later when Game Theory was established by the collaboration of mathematician John von Neumann and economist Oskar Morgenstern in their 1944 book *Game Theory and Economic Behavior*. In retrospect, the concept of Cournot equilibrium was regarded as an important predecessor of Nash equilibrium, a very key concept in mathematical Game Theory.

Quesnay, Cournot, and Walras were all noted French economists of world fame. There should be no objection against such a selection. The question that would naturally occur in our mind after listening to Schumpeter's remark on the Four Greatest Economists in History was like this. Who on earth was the last candidate Schumpeter really wanted to nominate as one of the Greatest Economists? Undoubtedly, another Frenchman should be ruled out. Someone would say that the right candidate would be Adam Smith, David Ricardo, John M. Keynes, or Karl Marx. A cynical person would dare to say that Schumpeter himself belongs to the Honorable Economist Club.

[1] For this point, see Yasui (1979). Takuma Yasui was one of the pioneers who tried his best to build a bridge between the old traditional economics and the modern economic science for his long years. A brief encounter with him at Osaka was an unforgettable memory in my long academic career.

We are ready to present another episode concerning the relationship between Cournot and Schumpeter. This is also an old real story that was made when Mr. Ichiro Nakayama, then a rising star from Japan, did research in economic theory at the University of Bonn, under the direction of Joseph Schumpeter. One day, old Schumpeter asked young Nakayama, "How have you been studying economics at a Japanese university?" Then, Nakayama honestly replied to Schumpeter "Yes, Sir. I myself was a humble assistant under the guidance of Professor Tokuzo Fukuda at Tokyo Shodai, who strongly recommended me to read Antoine Cournot, Heinlich Gossen, and Leon Walras." This was evidently a nice surprise to Schumpeter, who had a hard time swallowing the fact that the honorable names of Cournot, Gossen, and Walras were already known to Japan, a rather small island nation in the Pacific.[2]

In short, Schumpeter's appreciation of Cournot as a pioneering theorist was very high. Schumpeter's preference to equilibrium concepts in economic science was so strong that he selected Quesnay, Cournot, and Walras and the fourth unnamed person as the Honorable Club of the Four Greatest Economists in History. Although Cournot's own concept of equilibrium, characterized as "Cournot equilibrium," had been under constant attack in so many years, such a groundless attack quickly disappeared when the concept of "Nash equilibrium" as a natural successor of "Cournot Equilibrium" began to sweep over Game Theory and Applications.

In passing, it should be noticed that John Hicks, one of most influential economic theorists after Cournot, Walras, and perhaps Schumpeter, also appreciated the accomplishment of Cournot very much. For instance, in his 1935 paper on the theory of monopoly, Hicks (1935) remarked as follows:

> The generally increased interest in mathematical economics during the last few years has naturally turned attention back to the work of Cournot, the great founder of the subject, and still one of best teachers. It was Cournot's creation of elementary monopoly theory which was the first great triumph of mathematical economics; yet Cournot had left much undone. It is not surprising that the endeavor to complete his work has been an attractive occupation for his successors. (Hicks 1935, p. 1)

We agree with Hicks that Cournot should be thought of as the great founder of the theory of monopoly, duopoly, and oligopoly. Remarkably, Cournot had left much undone to his successors until the present time. For instance, following his pioneering work, there have been so many works done on the Theory of Oligopoly and Information in the 1980s, the 1990s, and even today. As the saying goes, life may be short but art (and science) should be quite long.[3]

[2]See Yasui (1979).

[3]For details, see Yasui (1979).

3.1.2 *John F. Nash as a Successor of Antoine Cournot*

John F. Nash (1928–2015) was an eccentric American mathematician with strong economic orientation. He made outstanding contributions to game theory, differential geometry, partial differential equations, and many other related fields. He shared the 1994 Nobel Memorial Prize in Economic Sciences with game theorists Reinhard Selten and John Harsanyi. He was born as a mathematical kid and received his Ph.D. at the age of only 22 years, with a mere 28-page dissertation on non-zero-sum games.

As the saying goes, there should be no clear lines between genius and madness. Already in 1959, Nash's mental illness began to come out in the form of paranoia, followed by more serious schizophrenia. While he was admired as a man with "beautiful mind" by his beautiful wife, he was also known as "the Phantom of Fine Hall" at Princeton University. In this chapter, regardless of Nash's personal character, we would like to just limit our attention to his outstanding contributions on non-zero-sum games and applications.[4]

We will briefly show that Cournot equilibrium is excellently succeeded by Nash equilibrium. To this end, let us first assume that there are two firms that produce a homogeneous product. Let x_i be the output of firm i ($i = 1, 2$), and let p be its common market price ($i = 1, 2$). Besides, let the inverse demand function be defined by $p = G(x_1 + x_2)$. If we define constant cost by c, then we can respectively define firm 1's profit and firm 2's profit as $\prod_1 = p\,x_1 - c\,x_1$ and $\prod_2 = p\,x_2 - c\,x_2$. Then, we say that the output pair (x_1*, x_2*) is a Cournot equilibrium pair if and only if the following conditions are both satisfied:

(C1) $\prod_1 (x_1{}^*, x_2{}^*) \geqq \prod_1 (x_1, x_2{}^*)$ for any x_1,

(C2) $\prod_2 (x_1{}^*, x_2{}^*) \geqq \prod_2 (x_1{}^*, x_2)$ for any x_2.

In plain English, the output pair ($x_1{}^*, x_2{}^*$) represents a Cournot equilibrium if and only if $x_1{}^*$ maximizes firm 1's profit $\prod_1 (x_1, x_2{}^*)$ for any given $x_2{}^*$, and $x_2{}^*$ maximizes firm 2's profit $\prod_2 (x_1{}^*, x_2)$ for any given $x_1{}^*$. Once the Cournot equilibrium is reached, it would be not a good policy at all for each firm to unilaterally change its output strategy. As a kind advice may tell us, please stay where you are!

Next, let us turn to a modern concept of Nash equilibrium in Game Theory. For simplicity, let us suppose that there are only two players, player 1 and player 2. Let S_i be the set of all possible strategies for player i ($i = 1, 2$). Let P_i be player i's payoff as a function of strategies, s_i. For the case of two players, we obtain $P_i = P_i (s_1, s_2)$ ($i = 1, 2$). Then, we say that the strategy pair ($s_1{}^*, s_2{}^*$) is a Nash equilibrium pair if and only if the following conditions are both satisfied:

(N1) $P_1 (s_1{}^*, s_2{}^*) \geqq P_1 (s_1, s_2{}^*)$ for any s_1,

(N2) $P_2 (s_1{}^*, s_2{}^*) \geqq P_2 (s_1{}^*, s_2)$ for any s_2.

Thus, the strategy pair ($s_1{}^*, s_2{}^*$) represents a Nash equilibrium if and only if $s_1{}^*$ maximizes player 1's payoff $P_1 (s_1, s_2{}^*)$ for any given $s_2{}^*$, and $s_2{}^*$ maximizes player

[4]For details, see Sakai (1990a, b, 1991), and Sakai and Sasaki (2021). For the personality and work of Nash, see Kuhn and Nasar (2002) and Nasar (1998).

2's payoff P_2 (s_1*, s_2) for any given s_1*. Once the Nash equilibrium is reached, it would be not a good policy at all for each player to unilaterally change its payoff strategy. Here again, please stay where you are![5]

If we compare a pair of Cournot equilibrium conditions, (C1) and (C2), and a pair of Nash equilibrium conditions, (N1) and (N2), then it is unquestionably clear that those two pairs of equilibrium conditions are perfectly analogous. We can naturally declare that the idea of Cournot has safely been succeeded by that of Nash. It is really one of "The Seven Wonders of the Academic World," however, that such succession from Cournot to Nash has been neglected for so many years. Here, we confirm again the following maxim. "Life is so short but art (or science) is so long."

3.2 The Residence Game as a Non-zero Sum Game: The Old Couple Versus the Young Couple

In what follows, we will carefully discuss the "Residence Game" as a typical example of a non-zero-sum game. Needless to say, living at an appropriate place together with good human relations constitutes the most important factor for our life. Besides, what we may call the "generation gaps problem" is closely related to such residence game. Let us suppose there are two couples the Old Couple and the Young Couple who are eager to find a new residence in a couple of months. It can be naturally expected to see that these two couple have similar or different opinions as to where to live and how they live. For instance, the Old Couple want to live in the Country with beautiful natural environments and hopefully in the neighborhood of the New Couple. While the New Couple may instead prefer to live in the City, in which housing facilities are modern and transportation means are well developed, they wish to live separately from and independently of their parents. As Russian novelist Ivan S. Turgenev (1818–1883) splendidly described in his famous novel *Fathers and Sons* (Turgenev 1862), the Old Generation may easily clash with the Young Generation over an even trifling problem, possibly causing a matter of life and death.[6]

Let us suppose that we send a questionnaire on residence preference to the two couples. The questionnaire asks a respondent about his/her choice of living in the country or living in the city and of living together (or living closely in the neighborhood) or living separately (or living far away) from relatives. On the one hand, we are informed that the Old Couple's first choice is to live in the country, and its

[5]For the life and work of Nash, see Nasar (1998). For the birth and development of game theory, see von Neumann and Morgenstern (1944), Suzuki (1999) and Varoufakis (2001). Also see Okada (2011).

[6]When I was a freshman at Kobe University, Japan, I read Turgenev (1862) for the first time. Although his masterpiece *Fathers and Sons* was published almost 100 years ago, it made a strong impression on me. I believe that the delicate and complicated relationship between the old and young generations may fundamentally be non-solvable and will continue forever.

Table 3.1 Rating results of the questionnaire I

Payoffs	Old couple	Young couple
10 points	Living in the country and together	Living in the city and separately
5 points	Living in the country and separately	Living in the city and together
3 points	Living in the city and together	Living in the country and separately
1 point	Living in the city and separately	Living in the country and together

		Young Couple: Country	Young Couple: City
Old Couple	**Country**	(living together) 10, 1	(living separately) 5, 10
	City	(living separately) 1, 3	(living together) 3, 5

Fig. 3.1 Residence Game I as the starting point. The pair (5, 10) represents only one solution

second choice is to live together with sons or daughters. On the other hand, however, we are informed that the Young Couple's first choice is to live in the city, and their second choice is to live separately from old parents. For simplicity, as is seen in Table 3.1, let us assume four different scores: 10 points, 5 points, 3 points, and 1 point.

As is clear in Table 3.1, a generations-gap problem is quite severe between the Old Couple and the Young Couple, so that finding a solution seems to be quite difficult. Between the two couples, both the first and second choices are entirely different. Under such circumstances, we will show that relying on the Nash equilibrium is so helpful for clear understanding. As the saying goes, seeing is believing!

We are now in a position to set up a sort of "Residence Game" as is shown in Fig. 3.1. There are two players in the game. They are "the Old Couple" and "the Young Couple." Each player, the Old Couple or the Young Couple, has the two possible strategies: namely, "living in the Country" and "living in the City." It is noted that each box in the matrix has a pair of numbers, where the first and second numbers respectively denote "the payoff of the Old Couple" and "the payoff of the New Couple." For instance, the pair (10, 1) in the upper left box indicates that when the both couples agree to live together (or at least to live in the neighborhood), the Old Couple gains 10 units of payoff and the Young Couple 1 unit of payoff. The pair (5, 10) in upper right box shows that when the Old Couple lives in the Country and the Young Couple in the City, the payoffs gained by each couple are respectively 5 units and 10 units. Similar interpretations can be done for the pair (1, 3) in the lower left box and the pair (3, 5) in the lower right box.

If we take a careful look at Fig. 3.1, then we can learn that the pair (5, 10) in the upper right box stands for only one Nash Solution. (This pair is enclosed by a small rectangle.) Let us explain why this is so. Now, suppose that the Old Couple lives in

Table 3.2 Rating results of the questionnaire II

Payoffs	Old couple	Young couple
10 points	Living together and in the country	Living separately and in the city
5 points	Living together and in the city	Living separately and in the country
3 points	Living separately and in the country	Living together and in the city
1 point	Living separately and in the city	Living together and in the country

the country and the New Couple in the city. Then, we can show that neither the Old Couple nor the New Couple has incentives to do separate activities. On the one hand, if the Old Couple unilaterally moves from the country to the city, then their payoff must decrease from 5 points to 3 points. On the other hand, if the Young Couple dares to move from the city to the country, then their payoff has to dramatically decline from 10 points to 1 point. Therefore, each couple should not change its strategy unless an opponent changes residence strategy. In short, once Nash equilibrium is achieved, the best strategy each player can take is to just obey the following conservative maxim: "Stay where you are!"

Moreover, we can show that the remaining three pairs cannot be Nash equilibriums. For instance, let us pick up the pair (10, 1) in the north-east box. It is true that if the Old Couple moves from the country to the city, then their payoffs will decrease from 10 units to 1 unit. So, the Old Couple has no intention whatever to change their residence. The New Couple, however, would have a different opinion, for its change of residence from the country to the city would definitely increase its payoff from 1 unit to 10 units. For similar reasons, the pairs (1, 3) and (3, 5) would not be qualified for solutions.

In short, in the present Residence Game I, the pair (5, 10) would be a "tentative solution," if not a "final solution." As far as we believe in usefulness of Nash equilibrium, there would exist no other solutions whatever. We bear in mind, however, that the Nash solution would represent only one possible solution, not a "perfect solution for guaranteeing happiness of all members." In fact, although the pair (5, 10) would make the Young Couple very happy with 10 units, the Old Couple's happiness would not be so high with mere 5 units. In other words, the present generation-gap solution is barely supported by suppression of old parents.

Next, we will pay attention to some other types of residence games. Let us recall that in Residence Game I above, the Old Couple's first choice is to live in the country, and their second choice is to live together with children, whereas the Young Couple's first choice is to live in the city, and their second choice is to live separately from parents. If we interchange the first choice and the second choice, we are expected to get into a completely new situation. Now, in the Residence Game II, whereas the Old Couple's first choice is to live together with children, and their second choice is to live in the country, the New Couple's first choice is to live separately from parents, and their second choice is to live in the city. More specifically, we obtain the *new* questionnaire results as seen in Table 3.2, from which we can derive the *new* payoff matrix of this Residence Game II in Fig. 3.2.

		Young Couple	
		Country	**City**
Old Couple	**Country**	(living together) 10, 1	(living separately) 3, 10
	City	(living separately) 1, 5	(living together) 5, 3

Fig. 3.2 Residence Game II as the starting point. There are no solutions at all

Let us compare this payoff matrix of Residence Game II with the last payoff matrix of Residence Game I. Then, we will see that the two pairs of payoffs (10, 5) and (3, 5) *on* the diagonal are the same in the two matrices. The two pairs of payoffs *off* the diagonal, however, are a bit different: exactly, the locations of the two numbers, 5 and 3, are just interchanged. Remarkably, such an apparently small difference will cause a substantial difference on the existence of a solution.

To our surprise, in this new residence game, there are no Nash solutions at all! Whichever box we would start with in Residence Game II, either Parents or Children would have good incentives to change their strategies. For instance, even if the pair (10, 1) in the upper-left box is initially chosen, the strategy change of Children from the country to the city would guarantee a big rise in their payoff from 1 to 10. Similarly, the pair (1, 5) cannot represent a Nash solution because the strategy change of Parents from the City to the Country would surely increase their payoff from 1 to 10. Similar interpretation would be possible if the remaining pairs are chosen as starting points.

Summing up, if Parents and Children play Residence Game II, the generations gap between them would be too huge to be reconcilable. Parents are not happy, nor are Children. So, this would be the worst situation we could think of.

Now, we are in a position to adequately combine Residence Games I and II, thus attempting to take a "middle position." While the Old Couple's choice is firstly to live together with Children and secondly to live in the country, the Young Couple's choice is firstly to live in the city and secondly to live separately from Parents. Presumably, such "crossing of preferences between Parents and Children" seems to frequently occur in today's Japan.

In this Residence Game III, the questionnaire results are summarized in Table 3.3, with the corresponding payoff matrix being shown in Fig. 3.3. As can easily be seen in Fig. 3.3, there should exist only one possible solution, which is represented by the pair (5, 5) in the lower-right box. In fact, once Parents and Children are "cornered" into this box, they should have no incentives to unilaterally change their strategies. Since this pair guarantees only a "second best solution" for Parents and Children, they have to feel "fairly satisfied" rather than "very happy." In fact, under such "middle situations," only each couple's first choice is realized, but their second choice is neglected. At any rate, although the generations-gap problem may happen

Table 3.3 Rating results of the questionnaire III

Payoffs	Old couple	Young couple
10 points	Living together and in the country	Living in the city and separately
5 points	Living together and in the city	Living in the city and together
3 points	Living separately and in the country	Living in the country and separately
1 point	Living separately and in the city	Living in the country and together

		Young Couple	
		Country	**City**
Old Couple	**Country**	(living together) 10, 1	(living separately) 3, 10
	City	(living separately) 1, 3	(living together) 5, 5

Fig. 3.3 Residence Game III. The only one possible solution (5, 5) makes Parents and Children "fairly satisfied," *but not* "very happy"

Table 3.4 Rating results of the questionnaire IV

Payoffs	Old couple	Young couple
10 points	Living together and in the country	Living together and in the city
5 points	Living together and in the city	Living together and in the country
3 points	Living separately and in the country	Living separately and in the city
1 point	Living separately and in the city	Living separately and in the country

at any time and in any country, it would really be a very tough problem to perfectly solve.

Now, let us turn our attention to the fourth possibility of multiple equilibrium. We are now concerned with the more friendly situation under which Parents and Children share the same residence option of living together. In spite of similarity of the first options, however, their second option is entirely different. Whereas Parents want to live in the peaceful country, Children wish to live in the convenient city. Then, we can obtain the *new* questionnaire results as seen in Table 3.4, from which we immediately derive the *new* payoff matrix of this Residence Game IV (Fig. 3.4).

As is quite clear in Fig. 3.4, there are *two* solutions. They are the pair (10, 5) in the upper left box, and the pair (5, 10) in the lower right box. If Parents and Children live together, Parents want to live in the lovely country, but Children wish to live in the exciting city. Unquestionably, it should be "mission impossible" to realize both Parents' wish and Children's wish simultaneously.

		Young Couple: Country	Young Couple: City
Old Couple	Country	(living together) 10, 5	(living separately) 3, 1
	City	(living separately) 1, 3	(living together) 5, 10

Fig. 3.4 Residence Game IV. When there are two Nash equilibriums, i.e., (5, 1) and (1, 5), a *third compromise plan* would be on the agenda

When there are *two* solutions. we are caught in a dilemma. If the Parents' wish comes first, the Children's wish is unduly ignored. And vice versa. Given two solutions, we cannot say which one is better than the other. If such a dilemma is to be dissolved, there should be no way out other than *rolling the dice*. Between the two players, we would imagine the friendly contract by which "an odd number" (1 or 3 or 5) gives Parents' preference a priority, but "an even number" (2 or 4 or 6) gives Children's preference a priority. Perhaps, such agreement would remind the reader of the *mixed strategies* introduced by von Neumann and Morgenstern for the solution of zero-sum games.

In order to get out of the dilemma, rolling the dice between Parents and Children would look like a good idea. A deeper reflection, however, would tell us that this is not a practical idea whatever, for interchanging their homes periodically would be very troublesome, thus giving much inconveniences to each player.

According to the Classical Theory of Yasuma Takata (1959), it is the Relative Strength of the Powers of the two players that finally determines where to live. If Parents have a stronger power than Children, then they would have to agree to live in the country. If Children succeed in persuading Parents, however, their choice of living in the country would prevail. Alternatively, the "*kuuki* or invisible social pressure" a la Shichihei Yamamoto (1997) would possibly prevail in the whole society, thus deeply influencing individual decisions. We believe that such sociological and historical considerations are very important but have been underestimated so far.[7]

[7] In Yamamoto (1997), Shichihei Yamamoto (1921–1991) intensively studies decision-making under *kuuki* or invisible social pressure, which differs from a Western way of individualistic decision-making based on rationality and consistency. During the Second World War, many *suicidal kamikaze attacks* by Japanese small zero-fighter planes were made against American big aircraft carriers and destroyers. Although such attacks appeared to be non-rational and even suicidal to many American people, they were highly praised by some patriotic Japanese people in those days. Such non-rational decision-making under social pressure still remains, perhaps to a lesser degree, in Japan today.

		BETTY Boxing	BETTY Ballet
JACK	Boxing	[5, 1]	−1, −1
JACK	Ballet	−1, −1	[1, 5]

Fig. 3.5 The lover's dilemma for a date

3.3 Some Other Non-zero-sum Games Reexamined

The modern theory of games has been closely related to the development of modern theory of economics. In this connection, there are a number of games that have drawn much attention. In what follows, we will critically reexamine several popular games and their applications.

3.3.1 "The Battle of the Sexes" or Plainly "The Lover's Dilemma for a Date"

The first popular game to discuss below has a very exciting name "the battle of the sexes," which literally means that the male sex and the female sex engage in a serious battle. We do not know, however, how serious the battle is. Since such naming sounds a bit exaggeration to us, we propose that it should be changed to a milder name "the lover's dilemma."

Let us suppose that there are two players, Jack and Betty. They are a pair of lovers who have different hobbies but are looking for a nice dating place. While Jack is very fond of seeing boxing, Betty likes to see ballet. Although each player wishes to see sport games with his/her partner, it is not an easy job for both players to reach an agreement. The worst scenario would be that they do not spend time together. More specifically, let us suppose that the payoff matrix of such a dating game can be expressed by Fig. 3.5. When a couple, Jack and Betty decide to see boxing, then Jack is very happy (5 points), but Betty feels so-so (1 point). In contrast to this, if they determine to see ballet, Betty is so happy (5 points), but Jack feels so-so (1 point). As a third possibility, we could imagine the situation under which Jack goes to see boxing and Betty chooses ballet, or Jack chooses ballet and Betty goes to see boxing. Then, they both would feel lonely and miserable (minus 1 point).

As far as the present date game is concerned, there exist two Nash equilibriums. They are the pair (Boxing, Boxing) and the pair (Ballet, Ballet). At the first equilibrium, Jack is very happy, but Betty is not so. At the second equilibrium, Betty feels so happy, but Jack does not so. Such non-symmetric situation may exaggeratedly be called "The Battle of the Sexes," but it should reasonably be

		BETTY		
		Boxing	* Movie	Ballet
JACK	Boxing	5, 1	−1, −1	−1, −1
	* Movie	−1, −1	* 3, 3	−1, −1
	Ballet	−1, −1	−1, −1	1, 5

Fig. 3.6 Introduction of a *third compromise option* of "going to a movie"

named "The Lover's Dilemma." The question that would immediately occur in our mind is in which way we can get out of this terrible dilemma.

One possible way out would be the adoption of "mixed strategies" by the two players, i.e., Jack and Betty. For instance, they could play the well-known game of Stone-Paper-Scissors to determine whose preference, Jacks' or Betty's, should be given the first priority. Careful observation of lovers would tell us, however, that this would be an unworkable idea. Presumably, a more reasonable idea would be that both Jack and Betty agree to seek a "second best solution." Without a certain compromise on both sides, a love relationship would sooner or later come to an end.

In order to get out of the dilemma, let us suppose that Jack and Betty wisely agree to adopt a "*third way* option of going to see Movie" instead of seeing either Boxing or Ballet. This alternative proposal would not make Jack and Betty very happy, but fairly acceptable.

Now, as is seen in Fig. 3.6, Jack and Betty will play the "Expanded Date Game," in which the *third strategy* "Movie" is newly added. Presumably, the strategy "Movie" will give each player only 3 units of payoffs. Then, the pair (Movie, Movie) = (2, 2) will be the third Nash solution. Understandably, this new solution would give less happiness to each player than the old solutions (Boxing, Boxing) and (Ballet, Ballet). They would nevertheless be satisfied with being together since they are true lovers.

The Game of Dating Lover's Dilemma has many applications to economic and related problems. Let us apply it to the Bargaining between of the Labor Union and the Manager Association. Whereas the Labor Union aims to improve the working conditions of union members such as increases in wages and pensions, the Manager Association wishes to increase the business conditions of the Company, by means of personnel reduction. Each player wants to have his/her demand approved by the opponent. Even if the opponent's reply is not perfectly satisfactory, it is the common knowledge among all players that the total breakdown should be avoided at any cost. In the light of these considerations, we can form the payoff matrix of the labor bargaining game as in Fig. 3.7.

As is seen in Fig. 3.7, there are *two* Nash solutions. The *first* solution (2, −1) shows the case of the worker's victory, but the *second* solution (−2, 3) means that of the manager's victory. In the real world, however, coexistence of two solutions may hardly be thinkable. In fact, the worker's position is generally weaker than the

		MANAGER	
		Wage Increase	Personnel Reduction
WORKER	Wage Increase	(settlement) 2, −1	(breakdown) −5, −3
	Personnel Reduction	(breakdown) −5, −3	(settlement) −2, 3

Fig. 3.7 The worker–manager bargaining game

		OTTO	
		Silent	Confess
ADOLF	Silent	(less charge for both) −2, −2	(9 yrs prison. set free) −9, −1
	Confess	(set free, 9 yrs prison) −1, −9	(6 yrs in prison for both) −6, −6

Fig. 3.8 The prisoner's dilemma

manager's position. As a result, a sort of "third compromise plan with a manager's advantage" is likely to be carried out. It is in such a situation that the famous "Theory of Power" a la Yasuma Takata (1959) could be highly effective.

3.3.2 The Prisoner's Dilemma

We will turn to the next problem of the prisoner's dilemma. After this problem was skillfully formalized by game theorist Albert W. Tucker, its popularity has constantly increased until today, with many economic and related applications.

Let us suppose that one day, after a bank robbery took place in the city, two persons, Adolf and Otto, were arrested as suspects. While they are now in the jail, each prisoner is in solitary confinement with no communications between them. Although the prosecutors do not have sufficient evidence to convict the pair on the serious charge of bank robbery, they obtain some evidence of convicting both on the minor charge of possession of illegal articles. So, the prosecutors have decided to offer each suspect a plea bargaining. Then, each suspect is given the chance either to unilaterally confess the pair's crime or to still keep silent. To sum up, as is seen in Fig. 3.8, we will have the have the following payoffs:

1. If both Adolf and Otto keep silent, then they will serve 2 years in prison.
2. If both Adolf and Otto confess the pair's crime, then they will serve 6 years in prison.
3. If Adolf (or Otto) keeps silent but Otto (Adorf) confesses the crime, Adolf (or Otto) will have to serve 9-long years in prison, but Otto (or Adorf) will be set free.

In this prisoner's dilemma game, there exists only one solution (Confess, Confess) = (−6, −6). This surely represents a Nash equilibrium, for a unilateral different action by one person would damage his position badly. The golden rule here is to "just stay where you are."

To tell the truth, the solution (−6, −6) is not only a Nash solution, but also a "stronger solution" or a "dominant strategy solution." In order to understand this, let us first assume that opponent Otto keeps silent. Then, Adolf 's payoff is (−2) if he also keeps silent, and 0 if he makes a confession, which shows that when Otto keeps silent, Adolf's best strategy is to confess a crime. Next, let us assume that Otto makes a confession. Then, Adolf 's payoff is (−1) if he keeps silent, and (−6) if he makes a confession, which shows that when Otto makes a confession, Adolf's best strategy is also to confess a crime. Therefore, whether Otto keeps silent or makes a confession, Adolf's best strategy is to confess a crime. In other words, to make a confession is Adolf's "dominant strategy." Similarly, to confess is also Otto's "dominant strategy." In short, the pair (Confess, Confess) represents a "dominant strategy equilibrium."

Now, we are ready to compare the following two pairs; (Silent, Silent) = (−2, −2) and (Confess, Confess) = (−6, −6). Clearly, for the two persons, the strategy "Silent" is better than the strategy "Confess." If communication between both was allowable, then each suspect would make the following promise between both; "I will keep silent whenever you keep silent. Please keep the promise." Such exchange of private opinions, however, is not possible because each prisoner is in solitary confinement with no communications allowable.

To summarize, when we are engaged in the prisoner's dilemma, we can recognize a serious gap between "individual rationality" and "whole rationality." In the prisoner's dilemma aforementioned, whereas the pair (Confess, Confess) seems to represent a sort of rationality, it does not produce the maximum level of the whole group. Each member's interest at the "micro level" does not necessarily correspond to the group interest at the "macro level." Micro is Micro, and Macro is Macro. Those two do not always go in the same direction.

The present prisoner's dilemma has many politico-economic applications. One of good examples is concerned with the "arms race." Let us recall the famous 1940 Charlie Chaplin Movie *The Great Dictator*, in which Hynkel as the dictator of Tomania and Napoloni as the dictator of Bacteria are fiercely involved in arms race negotiations for the world hegemony. As is seen in Fig. 3.9, the two rival countries have the two strategies: arms reduction and arms expansion.

There are several possibilities. If Hynkel and Napoloni maintain friendly relations and succeed in arms reduction negotiations, then they can utilize a part of military

		NAPOLONI: Reduction	NAPOLONI: Expansion
HYNKEL	Reduction	(friendly) 1, 1	(unilateral) —3, 2
	Expansion	(unilateral) 2, —3	(hostile) —1, —1

Fig. 3.9 The arms race: friendly, hostile, or unilateral

budget to the economy and welfare (plus 1 unit for payoff). If both countries are in hostile relations and decide arms expansion, then they will have to use more money for military budget (minus 1 unit). If Hynkel (or Napoloni) unilaterally pushes arms reduction, then the power of Tomania (or that of Bacteria) will decrease (minus 3 units) and the power of Bacteria (or that of Tomania) will increase (plus 3 units).

Clearly, a sort of the prisoner's dilemma takes place in the arms race. For Tomania and Bacteria, arms expansion is a dominant strategy, so that the pair (Expansion, Expansion) = $(-1, -1)$ represents a dominant strategy equilibrium. This equilibrium point is not only inferior to the pair (Reduction, Reduction) = $(1, 1)$, but also undesirable from global welfare. As far as Tomania and Bacteria do not trust each other, however, they will have to go on the way of arms expansion. As history tells us, human history is no more than a history of expansion, explosion, and destruction.

Another example of the prisoner's dilemma can be given by the "advertisement competition," which seems very excessive in reality. The basic framework of the advertisement competition has exactly the same as that of the arms race. We should now carry out a skillful reinterpretation task. Specifically, the two players Hynkel and Napoloni should be replaced by Tsukuba Company and Fuji Company, respectively. Moreover, the two strategies "Reduction" and "Expansion" should be reinterpreted as "Advertisement Reduction" and "Advertisement Expansion," but not as "Arms Reduction" and "Arms Expansion." Then, following the same analogy as before, we can easily show that the pair (Ad Expansion, Ad Expansion) is a dominant strategy equilibrium, thus the resulting competition between Tsukuba and Fuji Companies will be fierce and even excessive. Needless to say, excessive competition will yield the waist of financial and human resources. As the saying goes, doing too much is like doing too little.

3.4 Game Theory and Its Possible Faults: "Econs" Versus "Humans"

Soseki Natsume (1867–1916) is one of the most popular writers in Modern Japan. In the first page of his well-known book *Kusa-makura, or Lying on the Grass*, Natsume (1906) once remarked:

> Going up a mountain path, I have been thinking of the world where we live. If you simply follow the dictates of reason, you will create unnecessary troubles. If you use a pole in the stream of big emotions, you will be swept away by the strong current. If you faithfully rely on strong passions, you will be unduly confined. In short, the world where we live is not an ideal place for living. (Natsume 1906, p. 1)

As Natsume tells us, we live in the world where so many people with so many personalities live together. Roughly speaking, there are *three factors* that control human nature. They are Rationality, Emotion, and Passion. If we firmly believe in the power of rationality on the basis of cost and benefit, we may often offend the feeling of other people. If we are influenced by the strong power of emotion, we may sometimes lose the sense of balance, thus being swept away from important common sense. If we are strong passion lovers, our behavior will tend to go from one extreme to another. In short, any person should have a balanced mixture of Rationality, Emotion, and Passion.

Let us remind the reader of the exact title of the great book by John von Neumann and Oskar Morgenstern, *Theory of Games and Economic Behavior*. Here, "economic behavior," but not "human behavior," is mainly the focus of a discussion in the traditional game theory. Needless to say, "economic behavior" should be closely related to Rationality, thus underestimating the influence of Emotion and Passion. It is in this sense that the traditional game theory has a fundamental defect in a system, which must be mended as soon as possible.

According to Richard H. Thaler (2015), once a Rochester classmate and later a Nobel Prize winner, many traditional models tend to utilize a fictional creature called "Econs." Econs are a kind of "rational fools" a la Amartia Sen (1987) in the sense that, as is seen in traditional game theory, they rationally choose strategies by optimizing their payoffs. In other words, among the Three Natures of Human Beings, only Rationality is unduly picked up, with negligence of Emotion and Passion.[8]

In sharp contrast to the above argument, "Humans" are no more than human beings, namely *Homo sapiens*, Humans do a lot of misbehaving, meaning that traditional economic models lead to a lot of wrong predictions. Humans are

---

[8]Richard H. Thaler was once one of my close friends at the University of Rochenster, which was proud of employing many mathematics-oriented professors such as Lionel W. McKenzie, Edward Zabel, Huge Rose, Ronald Jones, and James Friedman. It is really a wonderful accomplishment that Rochester has produced a number of non-mathematical yet highly academic economists including Thaler. In short, a lot of Econs and Humans once lived, and still live, in Rochester.

supposed to possess many non-rational feelings such as envy, hatred, optimism, pessimism, sympathy, antagonism, and the like.[9]

In the real world, not in the fictional one, we are Humans rather than Econs! It is high time for us to bravely go beyond traditional game theory with the aim of establishing a brand new theory of human behavior. We need the second von Neumann and the second Morgenstern!

References

Akerlof GA, Shiller RJ (2019) Animal spirits: how the human psychology drives the economy, and why it matters for global capitalism. Princeton University Press

Hicks J (1935) Annual survey of economic theory: the theory of monopoly. Econometrica 3(1): 1–20

Kuhn WK, Nasar S (eds) (2002) The essential John Nash. Princeton University Press

Nasar S (1998) A beautiful mind. Simon & Schuster, New York

Natsume S (1906) *Kusamakura*, or sleeping on the grass (in Japanese). Asahi Shinbun Publishers, Tokyo

Okada A (2011) Game theory; new edition (in Japanese). Yuhikaku Publishers, Tokyo

Sakai Y (1990a) Theory of oligopoly and information (in Japanese). Toyo Keizai Publishers, Tokyo

Sakai Y (1990b) Information sharing in oligopoly: overview and evaluation. Part I: Alternative models with a common risk. Keio Econ Stud 27(2):17–41

Sakai Y (1991) Information sharing in oligopoly: overview and evaluation. Part II: Private risks and oligopoly models. Keio Econ Stud 28(1):51–70

Sakai Y (1995) Games and strategies: applications of Nash solution to many problems (in Japanese). In: Sakai Y (ed.) Economic theory (in Japanese). Yuhikaku Publishers, Tokyo

Sakai Y (2010) Economic thought of risk and uncertainty (in Japanese). Minerva Publishers, Kyoto

Sakai (2019) J.M. Keynes versus F.H. Knight: risk, probability, and uncertainty. Springer Nature, Singapore

Sakai Y, Sasaki K (2021) Information and distribution: the role of merchants in the market economy under uncertainty. Springer Nature, Singapore

Sen A (1987) On ethics and economics. Blackwell Publishing, Malden

Suzuki M (1999) The world of game theory (in Japanese). Keiso Publishers, Tokyo

Takata Y (1959) On power theory (in Japanese). Yuhikaku Publishers, Tokyo

Thaler RH (2015) Misbehaving; the making of behavioral economics. Norton, New York

Turgenev IS (1862) Fathers and sons. Grachev & Co., Moscow. [Russian original, many English translations]

Varoufakis Y (ed) (2001) Game theory: critical concepts in the social sciences. Routledge, London

Von Neumann J, Morgenstern O (1944) Theory of games and economic behavior. Princeton University Press

Yamamoto S (1997) Studies in *kuuki* or social pressure (in Japanese). Yamamoto Publishers, Tokyo

Yasui T (1979) My long life as an economist (in Japanese). Bokutaku Publishers, Tokyo

[9]For this point, see Akerlof and Shiller (2019) and Sakai (2010, 2019).

Part II
Decisions

Chapter 4
Consumer Decisions and Revealed Preference: Reevaluating Samuelson's Foundations of Economic Analysis

Abstract This chapter discusses consumer decision and revealed preference from many new angles, thus carefully reexamining the foundations of economic analysis. It is Paul A. Samuelson's contribution that boldly introduced to microeconomic theory the brand-new concept of *revealed preference* against the then standard doctrine of ordinal utility. The main result of this chapter is that Houthakker's *strong axiom of revealed preference* holds if and only if Samuelson's *weak axiom of revealed preference* and *a certain "regularity condition" newly introduced here* both hold. This new equivalence result distinguishes itself from all the previous works in the sense that no continuity assumptions are imposed on the demand function. In the light of uncertainty and behavioral economics in the present times, the significance and limitations of Samuelson-type economics are also discussed. Throughout this research, many graphical illustrations are presented for giving a helping hand to understand mathematical lemmas and theorems.

Keywords Revealed preference · Samuelson · The weak axiom · Houthakker · The strong axiom · Regularity conditions · Behavioral economics

4.1 The "Red Text" of the U.S.S.R. Institute Versus the "Blue Text" of Paul A. Samuelson

I have been doing research in economics for more than 60-long years. In hindsight, we recall that in the 1960s when we were ambitious students in Japan, there existed two popular yet opposing textbooks in economics. Remarkably, they were nicknamed the "Red Text" and the "Blue Text." The Red Text was well represented

---

This chapter is a completely revised version of Sakai (1973), with many new figures. I am grateful to Lionel W. McKenzie (Rochester), James W. Friedman (Rochester), William Brock (Chicago), Hiroshi Atsumi (Toronto), Trout Rader (Missouri), Edmond Malinvaud (Paris), and Werner Hildenbrand (Bonn) for kind suggestions. I am also thankful to Masashi Tajima (Shiga) for technical assistance.

Y. Sakai, *Games, Decisions, and Markets*, New Frontiers in Regional Science: Asian Perspectives 74, https://doi.org/10.1007/978-981-99-5285-4_4

by the U.S.S.R. Academy Economics Institute, Japan edition (1959) *Economic Textbook*, which was no doubt the most authoritarian text of the "Red Bloc" or the Soviet socialist bloc.[1]

The issue of "socialism versus capitalism" constituted the central theme of the red text. The honorable coauthors of the big text took pride in reaching the following provocative conclusions:

> We have thus far discussed the whole processes of economic development of a society. As a result, we have reached the most important conclusion that from a historical viewpoint, capitalism is destined to collapse whereas socialism is marching for its final victory over socialism. There should be no other way possible. We are so confident of such historical inevitability. (U.S.S.R. Academy Econ Institute, Japan ed. (1959), p. 1050)

In contrast to the only powerful Red Text, the Blue Text Group, which was regarded as a strong rival, seemed to be rather modest and even hesitant to the general public. Paul A. Samuelson, a representative of the Blue Text Group, wrote the world-popular textbook *Economics* (1955. 7th ed. 1967) and gently defended the American capitalist bloc in the following way:

> America leads Russia, but will the gap narrow? (Samuelson 1967, p. 791)

Comparing the "Red Text" and the "Blue Text," the majority of the people seemed to be more impressed by the powerful "Red Text" than the moderate "Blue Text." There existed the minority group including myself, however, who thought that the dominance of the "Red Wave" would be only temporary, and eventually overtaken by the "Blue Wave", which would behave like the "Great Wave" of famous Japanese *ukiyo-e* painter Hokusai.

Let us recall the street demonstrations and campus boycotting, which greatly hindered the Japanese students in their study. After some hesitation, I decided to go to the United States so that I could continue my graduate study without unduly political and psychological interruptions. I just wanted to get out of the Japanese university disturbance in the 1960's, thus daring to jump into the very core of the capitalist economy.

At appearance, it looked like an escape trip. However, in reality, it was an immense challenge or an enormous adventure to me. As the saying goes, God helps those who help themselves. In order to get financial and psychological help from the outside organizations, I had to do a lot of preparations for studies in advanced economics and higher mathematics.

---

[1] For the rivalry between capitalism and socialism, see Tsuru (1961) and Galbraith (1977).

4.2 Revealed Preference Theory: Paul A. Samuelson's Analysis as a Starting Point

Great book *Foundations of Economic Analysis* by Paul A. Samuelson (1947) served as a sort of bible for all the graduate students who already mastered more popular text *Economics* by Samuelson (1955, 7th ed. Samuelson 1967) and wanted to study at a more advanced level. The very first page per se was decorated with the following impressive passage:

> Mathematics is a Language. (Willard Gibbs, American mathematician (1839–1903))

I certainly guess that the young and ambitious Samuelson was quite impressed by such a short yet powerful message by an influential mathematician when he decided to write his masterpiece *Foundations of Economic Analysis.* Not only that, Chap. 1 of this book begins with the following useful passage:

> The existence of analogies between central features of various theories implies the existence of a general theory which underlies the particular theories and unifies them with respect to those central features. This fundamental principle of generalization by abstraction was enunciated by the eminent American mathematician E. H. Moore more than thirty years ago. It is the purpose of the pages that follow to work out its implications for theoretical and applied economics. (Samuelson 1947, p. 3)

As was well stated by E. H. Moore (1862–1932), although there exist a variety of *particular* theories, there is a *general* theory, which underlies those particular theories and unifies them with respect to those central features. This generalization principle was clearly stated by the eminent mathematician Moore and later confirmed by the young lion Samuelson. When I opened the first page of Samuelson's *Foundations*, I myself was greatly shocked like a thunderbolt from a clear sky and firmly felt that this was the time to say goodbye to the infertile and unproductive controversy between the "Red Text" and the "Blue Text" mentioned above. Following Moor and Samuelson, I then felt that I must do everything in my power toward a "high ivory tower" of unifying particular theories.

Revealed preference theory was first pioneered by the young Samuelson in his short article in Samuelson 1938 and later fully developed in his main work *Foundations* in 1947. Before Samuelson came on the stage, the existing theories of consumer demand were based on the traditional idea that consumers made consumption decisions to maximize their utility levels. For details for this point, see R. G. D. Allen (1936), J. R. Hicks (1st ed. 1930, 2nd ed. 1946), and others.

Revealed preference theory initiated by Samuelson requires us to make a drastic change in our way of thinking. Instead of stating with utility functions or ordinal preferences to derive demand functions, we first pay our attention to consumer's demand behavior and later infer consumer's preferences on the basis of a certain set of "rational behavior assumptions." The question to ask is what the "rational behavior assumption" is all about. Although there may be several assumptions conceivable, Samuelson's choice was as simple and lucid as what we now call "the weak axiom of revealed preference."

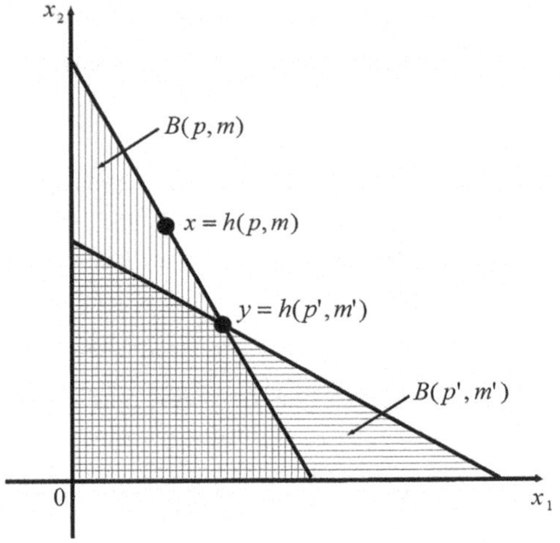

Fig. 4.1 Samuelson claims that x is "*revealed preferred*" to y. B (p, m) and B (p', m') are respectively indicated by the vertically shaded triangle and the horizontally shaded triangle

The fundamental idea of Samuelson's analysis is that "consumers reveal their preferences through their market behavior" if people's judgment is "rational and consistent." As is seen Fig. 4.1, let us suppose that there are two goods in the market, x_1 and x_2, and two bundles of goods, $x = (x_1, x_2)$ and $\mathbf{y} = (y_1, y_2)$. Let us also suppose that the two budget sets, B (p, m) and B (p', m'), and the two isoquants or utility curves, u and u', are drawn as nicely as in Fig. 4.1.

If the commodity bundles x and y are respectively chosen at the budget sets B (p, m) and B (p', m') by the consumer, we may conveniently obtain the following relations:

$$x = h\ (p, m) \in B\ (p, m), \tag{4.1}$$

$$y = h\ (p', m') \in B\ (p', m'). \tag{4.2}$$

It is clear in Fig. 4.1 that while $x = (x_1, x_2)$ and $y = (y_1, x_2)$ belong to the same budget set, B (p, m) $= \{(x_1, x_2)$: $p_1x_1 + p_2x_2 \leqq m\}$, meaning that both bundles can be bought at (p, m), only x is actually chosen.

Under such circumstances, Samuelson characteristically believes that x is *revealed preferred* to y. Someone might think that this is a too strong statement to accept. Later developments of consumer demand theory, however, have more or less taken Samuelson's side, with a certain qualification. In fact, we now know that the *weak axiom* of revealed preference a la Samuelson is not strong enough to guarantee the derivation of utility functions. We need to have something more than the *weak axiom*: Actually, the *strong axiom* of revealed preference that was later proposed by the cool-headed Houthakker should be adopted to successfully derive utility functions.

In short, this is the brief essence of revealed preference theory, showing the starting point of Samuelson's unique analysis. More detailed and technical discussions will gradually be done in the following sections.

4.3 Some Technical Preparations

We are concerned with a consumer faced with the problem of choosing a commodity bundle subject to given prices and income. Let us assume that there are n commodities, labeled $i = 1, ..., n$. Therefore, we will work within the n-dimensional vector space $\boldsymbol{R}^n$.

The commodity space Ω, or the set of all bundles, is defined as follows:

$$\boldsymbol{\Omega} = \{x = (x_1, \ldots, x_n) : x \geqq 0\}. \tag{4.3}$$

Let $\boldsymbol{P}$ and $\boldsymbol{M}$ be the price space and the income space, respectively. We make the following assumptions:

$$\boldsymbol{P} = \{p = (p_1, \ldots, p_n) : p \geqq 0\}, \tag{4.4}$$

$$\boldsymbol{M} = \{m : m \geqq 0\}. \tag{4.5}$$

Now, we denote by $\boldsymbol{B}$ the family of all competitive budgets, where for each $(p, m) \in \boldsymbol{P} \times \boldsymbol{M}$, each budget is defined as follows.

$$b\,(p, m) = \{x \in \boldsymbol{\Omega} : p\,x \leqq m\}. \tag{4.6}$$

Let h be a single-valued demand function on $\boldsymbol{B}$, that is, a function, which to each $b\,(p, m) \in \boldsymbol{M}$, assigns a commodity bundle $x = h\,(b\,(p, m))$. It is supposed here that $h\,(b\,(p, m)) = 0$ for $m = 0$. We denote by $\boldsymbol{X}$ the range of the demand function h:

$$\boldsymbol{X} = \cup\,\{h\,(b\,(p, m)) : b\,(p, m) \in \boldsymbol{M}\}. \tag{4.7}$$

In what follows, we make the two assumptions:

(D1) $\boldsymbol{X}$ is a convex set in $\boldsymbol{\Omega}$.

(D2) h satisfies the budget equation $p\,h\,(p, m) = m$ for any $(p, m) \in \boldsymbol{P} \times \boldsymbol{M}$.

For convenience, for $h\,(b(p, m))$ and $b(p, m) \in \boldsymbol{B}$, we will also simply write $h(p, m)$ and $(p, m) \in \boldsymbol{B}$.

We are now in a position to define "preference relations in terms of h" as follows. To this end, suppose the following relations hold:

(SRP) There is $b(p, m) \in \boldsymbol{B}$ such that $x = h(p, m)$, $y \in b(p, m)$, and $x \neq y$.

Then, following Samuelson (1947), we say that x is directly revealed preferred to y, and we simply write $x\,Sy$. If we look back at Fig. 4.1, we may easily understand the graphical meaning of the relation **(SRP)**. Indeed, seeing is believing!

Fortunately or unfortunately, (SRP) is not only one of revealed preference relations conceivable; Indeed, Houthakker (1950) succeeded in extending the Samuelson revealed relation **(SRP)** in a very natural way. More specifically, let us consider a new preference relation specified as a long chain of **(SRP)**:

(HRP) For some finite sequence $u^1, \ldots, u^m$, we have $x\;S\;u^1\;S \ldots u^m\;Sy$.

Then, in line with Houthakker (1950), we say that x is *indirectly revealed preferred to* y, conveniently writing xHy. In other words, H is the transitive closure of S on $\boldsymbol{\Omega}$. Clearly, xSy implies xHy, but not the other way around.

Samuelson's weak axiom **(W)** and Houthakker's strong axiom **(S)** of revealed preference are formulated as follows:

(W) For any $x, y \in \boldsymbol{\Omega}$, $x\,Sy$ implies $\sim y\,Sx$.

(S) For any $x, y \in \boldsymbol{\Omega}$, $x\,Hy$ implies $\sim y\,Hx$.

In other words, the weak axiom **(W)** argues that the relation S is not symmetric, whereas the strong axiom **(S)** states that the relation H is not symmetric. Each axiom clearly shows the consistency of consumer's choice behavior in a direct or indirect way.

Let $\boldsymbol{E}\,(p)$ be the well-known Engel curve associated with a price vector p:

$$\boldsymbol{E}\,(p) = \{h\,(p, m) : m \in \boldsymbol{M}\}, p \in \boldsymbol{P}. \tag{4.8}$$

Denoting by $Cls(\boldsymbol{X})$ the closure of $\boldsymbol{X}$, let us define the following sets of commodity bundles:

$$\underline{\boldsymbol{A}}\,(x^0) = \{x \in Cls\,(\boldsymbol{X}) : \sim x\,H\,x^0\}, x^0 \in \boldsymbol{X}, \tag{4.9}$$

$$\bar{\boldsymbol{A}}\,(x^0) = \{x \in Cls\,(\boldsymbol{X}) : \sim x^0\,H\,x\}, x^0 \in \boldsymbol{X}. \tag{4.10}$$

On the one hand, the under-bar set $\underline{\boldsymbol{A}}(x^0)$ stands for the set of commodity bundles that are not indirectly revealed preferred to a given commodity bundle x^0. On the other hand, the upper-bar set $\bar{\boldsymbol{A}}(x^0)$ denotes the set of commodity bundles to which a given commodity bundle x^0 is not indirectly revealed preferred.

Now, in order to postulate a "regularity condition" under which the weak and strong axioms are to be established, it is quite useful for us to introduce two different kinds of *expenditure compensation functions* as follows[2]:

$$\tilde{\boldsymbol{a}}(p, x^0) = \sup\,\{p\,x : x \in \underline{\boldsymbol{A}}(x^0) \cap \boldsymbol{E}\,(p)\}, (p, x^0) \in \boldsymbol{P} \times \boldsymbol{X}, \tag{4.11}$$

$$\underline{\boldsymbol{a}}\,(p, x^0) = \inf\,\{p\,x : x \in \bar{\boldsymbol{A}}(x^0) \cap \boldsymbol{E}\,(p)\}, (p, x^0) \in \boldsymbol{P} \times \boldsymbol{X}. \tag{4.12}$$

On the one hand, the newly defined function $\tilde{\boldsymbol{a}}\,(p, x^0)$ stands for the supremum of those expenditures px $(=m)$ at price p such that the commodity bundle $x = h\,(p, m)$ is not indirectly revealed preferred to a given commodity bundle x^0. On the other hand, the function $\underline{\boldsymbol{a}}\,(p, x^0)$ indicates the infimum of those expenditures px (= m) at price p such that a given commodity bundle x^0 is not indirectly revealed preferred to the commodity bundle $x = h\,(p, m)$. It will be seen that under assumptions (D1), (D2), and (S), the functions $\tilde{\boldsymbol{a}}\,(p, x^0)$ and $\underline{\boldsymbol{a}}\,(p, x^0)$ are well defined for all $(p, x^0) \in \boldsymbol{P} \times \boldsymbol{X}$.

---

[2]The expenditure compensation function was first introduced and effectively employed by McKenzie (1957).

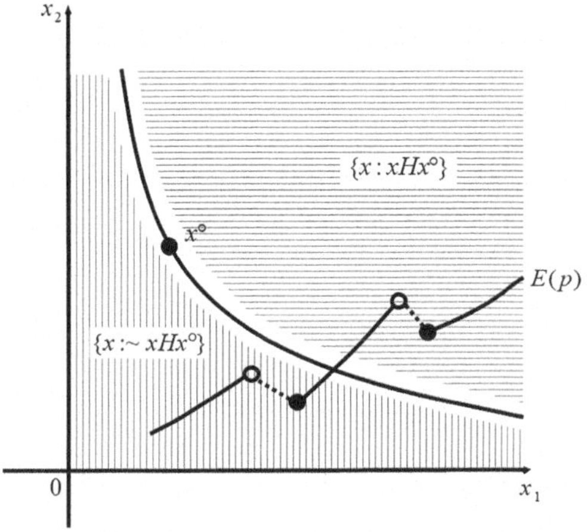

Fig. 4.2 The upper-shaded set {x: xHx^0} and the lower-shaded set {x: ~ xHx^0}. The Engel curve E (p) may, or may not, be continuous. It breaks twice here

At first glance, the meanings of those functions may appear to be complicated and difficult for the reader to comprehend. As the saying goes, however, seeing is really believing. Graphical representations will help us to easily understand what is going on here. For instance, in Fig. 4.2, the whole commodity set $\boldsymbol{X}$ is the union of the upper-shaded set {x: xHx^0} and the lower-shaded set {x: ~ xHx^0}, where the base commodity x^0 belongs to the lower-shaded set, but not the upper-shaded set. Depending on the condition of the demand curve h (p, m), the Engel curve $\boldsymbol{E}$ (p) may, or may not, be continuous. If it happens to be continuous, it can be drawn as a continuous curve like $\boldsymbol{E}$ (p) in Fig. 4.2. The expenditure compensation function $\tilde{\boldsymbol{a}}$ (p, x^0) can roughly be depicted as a downward-sloped line there.

We are now in a position to carefully introduce the *regularity condition*, which has to play a very important role as a bridge between the weak and strong axioms of revealed preference. Specifically, such a *regularity condition* is formulated as follows:

(R) For all (p, x^0) $\in \boldsymbol{P} \times X$, $\tilde{\boldsymbol{a}}$ (p, x^0) $\geqq \underline{\boldsymbol{a}}$ (p, x^0).

It would not be an easy job to describe the exact meaning of the condition (**R**). Broadly speaking, the condition (R) is that for a given price vector p and a given commodity bundle x^0, the maximum of "worse-than-or-as-good-as- x^0" expenditures is greater that or equal to the minimum of "better-than-or-as-good-as x^0" expenditures. Once again, graphical illustrations would help us to understand what is going on here. Figure 4.3 shows the situation under which the condition holds, while Fig. 4.4 shows the situation under which it is violated. As is clear from a comparison between those two figures, the regularity condition (**R**) requires that there be no "*~H gap*" on the Engel curve $\boldsymbol{E}$ (p) for any p, yet allowing for "*~H overlap" on* it. It should be noted that there may exist "jumps" on the Engel curve since the continuity of the demand function is not guaranteed here.

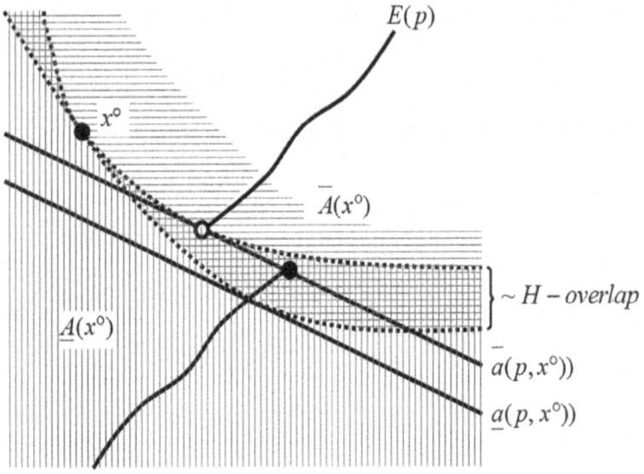

Fig. 4.3 $E(p)$ is not continuous, jumping once. Remarkably, (**R**) holds in this case

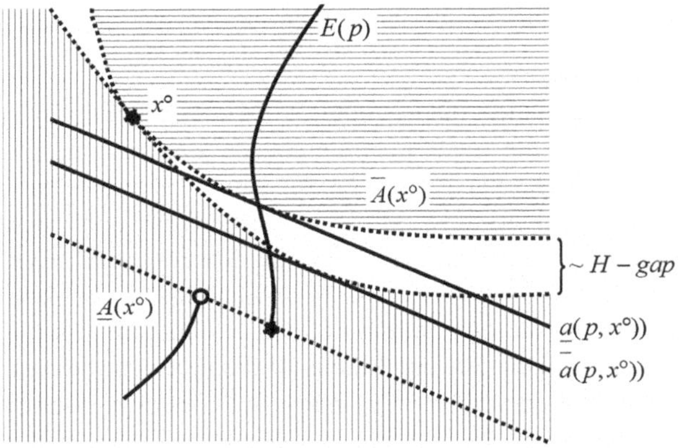

Fig. 4.4 (p) is not continuous, jumping once. And (**R**) is violated

4.4 Equivalence of the Weak and Strong Axioms of Revealed Preference

In this section, we would like to establish an important equivalence theorem of the weak and strong axioms of revealed preference on *Cls* (***X***), the closure of the range of the demand function. More exactly, we wish to demonstrate that, for the demand function satisfying **(D1)** and **(D2)**, the strong axiom (**S**) holds on *Cls* (***X***) if and only if the weak axiom (**W**) together with the regularity condition (**R**) holds on *Cls* (***X***). This is no doubt the core of our research, requiring a careful, step-by-step inquiry.

We need several preliminary results in preparation for the equivalence theorem. To this end, we first note the following lemma.

Lemma 4.1: $\bar{A}$ as a Closed Set Let demand function h satisfy **(D1, 2)** and the weak axiom (**W**). Then, for any $x^0 \in \boldsymbol{X}$, the upper-bar set $\bar{A}\ (x^0)$ is a closed set in *Cls* (***X***).

Proof It suffices to show that the set $\{x \in Cls\ (\boldsymbol{X}){:}\ x^0Hx\}$, $x^0 \in \boldsymbol{X}$, is an open set in *Cls* (***X***).

First, if $x^0 = 0$, then this set is evidently empty, so that it is open. Next, if $x^0 \neq 0$, then let us choose any $x^1 \in \boldsymbol{X}$ such that $x^0 H x^1$. Then, by the definition of H, there exists $x^2 \in \boldsymbol{X}$ such that

$$x^0 = x^2 \text{ or } x^0\, H\, x^2, \\ \text{and } p^2 x^2 \geqq p^2 x^1,\ x^2 \neq x^1. \tag{4.13}$$

Let $x^t = (1-t)x^1 + t\,x^2, 0 \leqq t \leqq 1$. Then, we find $x^t \in X$ since X is convex by (D1). So, we should have the following:

$$p^2\, x^2 \geqq p^2\, x^t, x^t \neq x^1. \tag{4.14}$$

This relation together with (**W**) must yield the following results:

$$p^t\, x^2 > p^t\, x^t. \tag{4.15}$$

Because of definition of x^t, we must have the following equation:

$$(1-t)\ (p^t\, x^1 - p^t x^t) + t\ (p^t\, x^2 - p^t x^t) = 0,$$

whence we find $p^t\, x^t > p^t\, x^1$ by Eq. (4.15). Here, we can choose a neighborhood $V\,(x^1)$ of x^1 such that

$$p^t\, x^t > p^t\, y \text{ for any } y \in V\ (x^1) \cap Cls\ (\boldsymbol{X}). \tag{4.16}$$

If we combine Eqs. (4.13), (4.14) and (4.16), then we obtain $x^0 Hy$ for any $y \in V\,(x^1) \cap Cls\,(\boldsymbol{X})$. This ensures the desired result that the set $\{x \in Cls\,(\boldsymbol{X}){:}\ x^0 Hx\}, x^0 \in \boldsymbol{X}$, is an open set in $Cls\,(\boldsymbol{X})$.

Q.E.D.

As is seen above, the proof of Lemma 4.3 is tedious but straightforward. Its graphical and intuitive outline, showing that the set $\{x{:}\ x^0 Hx\}$ is open, may be given in Fig. 4.5. This lemma will effectively be utilized in the equivalence theorem.

To reach the final goal of the equivalence theorem, we need to pass through the expenditure compensation functions to be stated below.

Lemma 4.2: $\tilde{\boldsymbol{a}}$ and $\underline{\boldsymbol{a}}$ Are Well Defined Let the demand function h satisfy **(D1, D2)** and the strong axiom (**S**). Then, for any $(p, x^0) \in \boldsymbol{P} \times X$, the expenditure compensation functions $\tilde{\boldsymbol{a}}\,(p, x^0)$ and $\underline{\boldsymbol{a}}\,(p, x^0)$ are well defined.

Proof[3] To see that the $\tilde{\boldsymbol{a}}\,(p, x^0)$ is well defined, it suffices to see the following set is non-empty and bounded above (see Fig. 4.6):

[3] I am grateful to Uzawa (1960) for providing a guide to the proof of this lemma. Its completely revised version was contained in Chipman et al. (1971).

Fig. 4.5 The set $\{x: x^0Hx\}$ is open. A graphical outline illustrated here

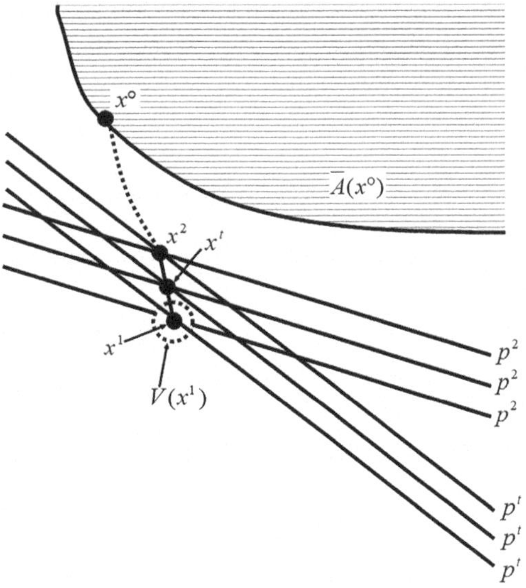

Fig. 4.6 The proof of Lemma 4.2: a graphical outline

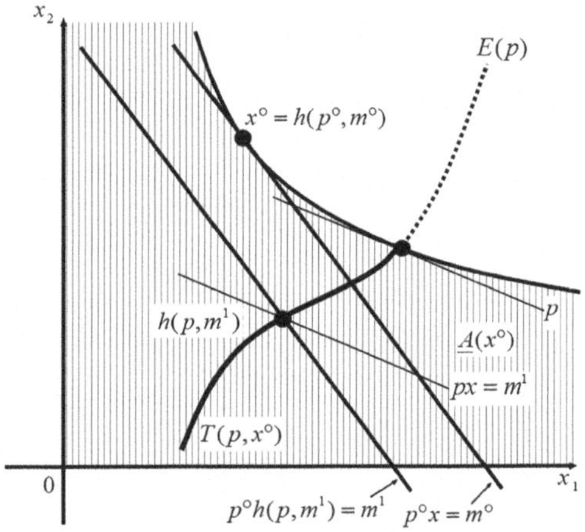

$$T\,(p, x^0) \equiv \{p\,x : x \in \underline{\boldsymbol{A}}(x^0) \cap \boldsymbol{E}(p)\,\}. \tag{4.17}$$

To this end, let us choose any $(p, x^0) \in \boldsymbol{P} \times \boldsymbol{X}$, and note $x^0 = h\,(p^0, m^0) \in \boldsymbol{B}$. There are two possibilities for the value of m^0: namely, $m^0 = 0$ or $m^0 > 0$. If $m^0 = 0$, then it is clear that $\boldsymbol{T}\,(p, x^0) = \{0\}$. If $m^0 > 0$, then the set $\{x: 0 \leqq x \in Cls\,(\boldsymbol{X})\ \&\ p^0x < m^0\}$ is nonempty. It is possible to find $m^1 \in \boldsymbol{M}$ such that, in $Cls\,(\boldsymbol{X})$, the hyperplane $px = m^1$ lies below the hyperplane $p^0x = m^0$. Therefore, we have $p^0h(p, m^1) < p^0x^0$. This implies $x^0Sh(p, m^1)$, so that $x^0Hh\,(p, m^1)$. It follows from (***S***) that we find $\sim h(p, m^1)\,Hx^0$, thus implying that $ph\,(p, m^1) \in \boldsymbol{T}\,(p, x^0)$. This ensures that for any $(p, x^0) \in \boldsymbol{P} \times \boldsymbol{X}$, the set $\boldsymbol{T}\,(p, x^0)$ is nonempty.

Next, for any $(p, x^0) \in \boldsymbol{P} \times \boldsymbol{X}$, let us choose $m^2 \in \boldsymbol{M}$ such that $px^0 < m^2$. It follows that $ph(p, m^2) = m^2 > px^0$, implying that $h(p, m^2)Hx^0$. This ensures that $ph(p, m^2)$ is an upper bound to the set $\boldsymbol{T}(p, x^0)$.

The proof that $\underline{\boldsymbol{a}}(p, x^0)$ is well defined for any $(p, x^0) \in \boldsymbol{P} \times \boldsymbol{X}$ proceeds in a similar way.

Q.E.D.

With all the preparations aforementioned, we are now in a position to state and prove the equivalence theorem, which certainly represents the most important result of this chapter.

Theorem 4.3: Equivalence Theorem Let the demand function h satisfy **(D1, D2)**. Then, the *strong axiom* **(S)** holds if and only if the *weak axiom* **(W)** and the *regularity condition* **(R)** both hold.

Proof

1. **(S)** ⇒ **(W)** & **(R)**[4]:

   Let $h(p, m)$ satisfy the *strong axiom* **(S)**. Then, by definition per se, the *weak axiom* **(W)** is obviously implied by (S). Note that, as was shown in Lemma 4.2, the expenditure compensation functions $\tilde{\boldsymbol{a}}(p, x^0)$ and $\underline{\boldsymbol{a}}(p, x^0)$ are well defined for any $(p, x^0) \in \boldsymbol{P} \times X$. In order to derive the *regularity condition* (R), we rely on proof by contradiction. To this end, let us dare to suppose to the contrary that $\tilde{\boldsymbol{a}}(p, x^0) < \underline{\boldsymbol{a}}(p, x^0)$ for some $(p, x^0) \in \boldsymbol{P} \times X$. Then, choose $m \in M$ so that $\tilde{\boldsymbol{a}}(p, x^0) < m < \underline{\boldsymbol{a}}(p, x^0)$, and let $x = h(p, m)$. Then, we immediately obtain the following:

   $$\tilde{\boldsymbol{a}}(p, x^0) < p\,x < \underline{\boldsymbol{a}}(p, x^0). \tag{4.18}$$

   By the definition of ã and $\underline{a}$, we obtain the following:

   $$x\,H\,x^{\mathbf{0}} \text{ and } x^0\,H\,x, \tag{4.19}$$

   which clearly contradicts the *strong axiom* **(S)**. So, in order to get rid of the contradiction, we should have $\tilde{\boldsymbol{a}}(p, x^0) \geqq \underline{\boldsymbol{a}}(p, x^0)$, which is nothing but the *regularity condition* **(R)**.

2. **(W)** & **(R)** ⇒ **(S)**:

   Next, we will prove that the *weak axiom* **(W)** together with the *regularity condition* (R) implies the *strong axiom* (S). Once again, we rely on the proof by contradiction. So, let us assume otherwise. Then, there exist some commodities $x, x^0 \in X$ such that the following relations hold:

[4]I am indebted to Lionel W. McKenzie for suggesting the present proof, which was better than my previous proof.

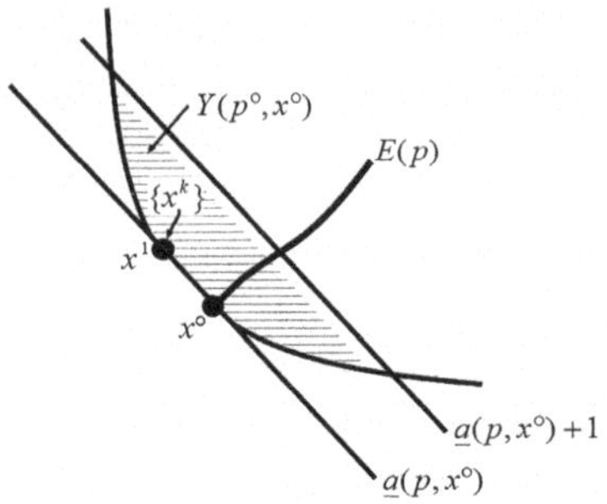

Fig. 4.7 The set $Y(p, x^0)$ in the proof of Theorem 4.3 is illustrated here

$$x\, H\, x^0 \text{ and } x^0\, H\, x. \tag{4.20}$$

Let $x = h\,(p, m)$ for some $(p, m) \in \boldsymbol{B}$. Then, we will show that Eq. (4.20) implies the following:

$$\tilde{\boldsymbol{a}}\,(p, x^0) \leqq p\,x \leqq \underline{\boldsymbol{a}}\,(p, x^0). \tag{4.21}$$

Indeed, if we assume on the contrary that $\tilde{\boldsymbol{a}}\,(p, x^0) > p\,x$, we could take an $x^* \in \underline{\boldsymbol{A}}(x^0) \cap \boldsymbol{E}\,(p)$ for which $\tilde{\boldsymbol{a}}\,(p, x^0) \geqq px^* > p\,x$. The last inequality yields $x^*\, S\, x$, which together with xHx^0 entails x^*Hx^0. This contradicts $x^* \in \underline{\boldsymbol{A}}(x^0)$. In order to get rid of the contradiction, we should originally have $\tilde{\boldsymbol{a}}\,(p, x^0) \leqq px$.

The proof that x^0Hx with $x \in \boldsymbol{E}\,(p)$ implies $p\,x \leqq \underline{\boldsymbol{a}}\,(p, x^0)$ proceeds in an analogous way.

Now, let us specifically assume $px = m = \underline{\boldsymbol{a}}\,(p, x^0)$. Then, take a sequence $\{p\,x^{\,i}\}$ converging to $p\,x$ such that $x^{\,i} \in \bar{\boldsymbol{A}}(x^0) \cap \boldsymbol{E}\,(p)$. This is possible by the definition of $\underline{a}$. Define the new set $\boldsymbol{Y}\,(p^0, x^0)$ in the following way (see Fig. 4.7 for an illustration):

$$\boldsymbol{Y}\,(p^0, x^0) = \bar{\boldsymbol{A}}\,(x^0) \cap \left\{x \in Cls\,(X) : \underline{\boldsymbol{a}}\,(p, x^0) \leqq p\,x < \underline{\boldsymbol{a}}\,(p, x^0) + 1\right\}. \tag{4.22}$$

Let us focus on the sequence $\{x^i\}$ aforementioned. Since $x^i \in Y\,(p^0, x^0)$ for sufficiently large i, we may without loss of generality assume that the original sequence $\{x^i\}$ itself is a sequence in $\boldsymbol{Y}\,(p^0, x^0)$. Since $\bar{\boldsymbol{A}}(x^0)$ is a closed set in *Cls* (*X*) by means of Lemma 4.1, $\boldsymbol{Y}\,(p^0, x^0)$ is indeed a compact set in *Cls* (*X*). Therefore, we can take a convergent subsequence $\{x^k\} \subset \{x^i\}$ in $\boldsymbol{Y}\,(p^0, x^0)$. Let $x^1 = \mathbf{Lim}\, x^k \in \boldsymbol{Y}\,(p^0, x^0) \subset \bar{\boldsymbol{A}}(x^0)$. Then, we obtain $px^1 = px$. Here, note that either $x^1 = x$ or $x^1 \neq x$ since h may, or may not, be continuous. If $x^1 = x$, then by means of Eq. (4.20), we have x^0Hx^1, contradicting that $x^1 \in \bar{\boldsymbol{A}}(x^0)$. If $x^1 \neq$ x, then the equality $px = px^1$yields xSx^1, which together with Eq. (4.20) implies x^0Hx^1. This also contradicts that $x^i \in \bar{\boldsymbol{A}}(x^0)$. To get out of the contradiction, we cannot specifically assume $p\,x = \underline{\boldsymbol{a}}\,(p, x^0)$.

Hence, in the light of Eq. (4.21), we must have the following relations.

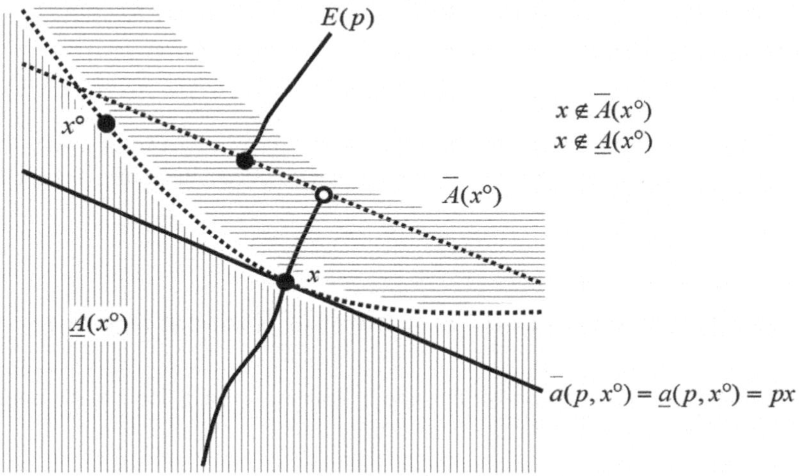

Fig. 4.8 The set $Y(p^0, x^0)$ is illustrated here for the proof of Theorem 4.3

$$\tilde{\boldsymbol{a}}\ (p, x^0) \leqq p\, x < \underline{\boldsymbol{a}}\ (p, x^0). \tag{4.23}$$

This clearly contradicts the regularity condition (**R**). And now, we can happily declare that the proof is complete.

Q.E.D.

Needless to say, Theorem 4.3 (the equivalence theorem) represents the most important result in this chapter. As is easily seen in the detailed proof aforementioned, Lemma 4.1, which demonstrates that the set $\bar{\boldsymbol{A}}\ (x^0)$ is closed under (**W**), plays a critical role in the proof part. Indeed, if the set $\bar{\boldsymbol{A}}\ (x^0)$ is NOT closed, we may have the case in which $x \notin \bar{\boldsymbol{A}}\ (x^0)$ and $x \notin \underline{\boldsymbol{A}}\ (x^0)$, meaning that x^0Hx and xHx^0, and $\tilde{\boldsymbol{a}}\ (p, x^0) = \underline{\boldsymbol{a}}\ (p, x^0) = px$, $x \in \boldsymbol{E}\ (p)$. For a graphical outline of this point, see Fig. 4.8**.** In such a case, h surely satisfies the regularity condition (**R**), but not the strong axiom (**S**).

It is worthwhile to discuss how our equivalence result distinguishes itself from the previous results in the related literature. First of all, we can say that the hypothesis of Theorem 4.3 is definitely weaker than that of van Moeseke's Theorem 2.5 (Fox et al. 1969, p. 99). This is because (1) neither Lipschitz nor continuity conditions are imposed on the demand function h, and (2) X need not be the positive or nonnegative quadrant of R^n: namely, it can be any convex subset.

Let us recall that Sonnenschein's Example 2 (Chipman et al. 1971, p.274) shows that (D1, D2) and (**S**) do not imply the continuity of h. In view of Theorem 4.3, it is easily seen that the regularity condition (**R**) [even together with (D1, D2) and (**S**)] does not suffice to guarantee the continuity of h. Besides, based upon the strong axiom of revealed preference (**S**), Hurwicz and Richter (1971) carefully studied revealed preference theory *without demand continuity assumptions.*[5]

In the light of these works mentioned above, the significance of our equivalence theorem *without demand continuity assumptions* is quite evident. Generally speaking, continuity assumptions may be useful yet strong in any subject of research. They may unfortunately be too strong in some research areas including the present consumer demand theory.

4.5 The Effectiveness and Limitations of Rational Decisions

In his famous book *Foundations of Economic Analysis* (1947), Paul A. Samuelson (1947) remarked as follows;

> I was aware that each field involved interdependent unknowns determined by presumably efficacious, interdependent equilibrium conditions—a fact which has always been generally realized. But, and this leads me to the second fundamental purpose of this work, it had not been pointed out to my knowledge that there exist formally identical *meaningful* theorems in these fields, each derived by an essentially analogous method. This is not surprising since only the smallest fraction of economic writings, theoretical and applied, has been concerned with the derivation of *operationally meaningful* theorems. (Samuelson 1947, p. 3)

The main purpose of Samuelson' s *Foundations* (1947) was to obtain a series of *operationally meaningful* theorems, which had unfortunately been very rare in the 1940s. In fact, when it published at long last after an unavoidable delay due to the Second World War, it received a sort of mixed blessing in the then dominant economics profession. For instance, Kenneth Boulding (1948), one of the finest contemporaries at that time, wrote the following rather critical and cynical review:

> The *Foundations* is an important book. It should be studied not only by the mathematically baptized but also by those who, like myself, hang on to n-dimensions by the skin of their teeth. No economist who studies it can fail to profit by it. Nevertheless, the present reviewer cannot help feeling a certain sense of rapidly diminishing marginal productivity. in the application of mathematics to economics. There is an elusive flavor of John Stuart Mill about the *Foundations* which makes it seem less like a foundation than a coping stone, finishing an edifice which does not have much further to go. It may well be that the slovenly literary borderland between economics and sociology will be the most fruitful building during the years to come and that mathematical economics will remain too flawless in its perfection to be very fruitful. (Boulding 1948, p. 199.)

As Boulding eloquently tells us, Samuelson's *Foundations* is an important book. Every economist would agree with this. Indeed, by making use of the weak axiom of

[5]When I established the equivalence theorem with no continuity assumptions, I received a very positive response from Edmond Malinvaud, one of the leading theoretical economists. In fact, in his famous advanced text, Malinvaud (1972, revised 1985) praised my early paper (1973) by saying that Sakai finished the equivalence theorem in consumer demand theory. It was certainly a nice surprise to me. Note that besides my regularity condition (**R**), there are several other regularity conditions conceivable (see Sakai (1973)).

revealed preference with help of his friend Houthakker, Samuelson succeeded in deriving many *meaningful theorems* in consumer choice theory. So far so good. There exist a certain number of economists, however, who might rather put themselves in Boulding's position. We should be very careful of Boulding's remark on Samuelson's *Foundations,* which never admires that *Foundations* is a *very important book*, only saying that it is just an important book. This descriptive difference looks small, but may sometimes be larger than we imagine.[6]

As far as I am concerned, my own position on *Foundations* is rather delicate. When I was young and gay, I found much interest in mathematical economics and read Samuelson's many books and papers repeatedly and energetically. To tell the truth, this paper is no more than a completely revised version of Chapter One of my Ph.D. thesis *Foundations of Consumption and Production Theories* submitted to the University of Rochester in September 1972. Professors Lionel W. McKenzie and James W. Friedman were my thesis advisers, thus giving me kind suggestions and constant encouragement. When I was later teaching economic theory at the University of Pittsburgh, my attitude on the application of mathematics to economics gradually changed into the direction of more economic reality rather than pure theory per se, with growing interest in building a bridge between Theory and History. Since John R. Hicks happens to be McKenzie's teacher at Oxford, I am a sort of Hicks' grand child, and then have followed the new Hicks (1969), who has found growing interest in *Theory of Economic History.*[7]

In hindsight, although Boulding's prediction that "the slovenly literary borderland between economics and sociology will be the most fruitful building during the years to come" may not be totally right, I nevertheless think that it partially hits the target. In fact, from the 1980s to the present 2020s, there have gradually yet surely emerged new waves of social sciences toward "more psychology, more sociology and still more history." Amartia Sen (1977, 1987) is one of those economists, who has constantly issued a strong warning against *excessively rational and unduly consistent human behavior*, for it would possibly lead us to the non-realistic world where all men and women as " rational fools" make neither mistakes nor contradictory behaviors. As the saying goes, extremely spicy things are not good for the stomach. Very noted Japanese essayist Shichihei Yamamoto (1983) wrote a very

[6] Before the review of Boulding (1948) appeared, the ordinal utility approach represented by Allen (1936) and Hicks (1939, revised 1946) dominated consumer demand theory for a long time. Although Hicks was influenced by the rise of revealed preference approach, he took an independent middle path between the two approaches. For such a subtle position, see Hicks (1956).

[7] It was in his later years that noted theoretician John Hicks (1969) published a very interesting book *A Theory of Economic History,* in which two apparently different subjects—Theory and History—were nicely combined into one. In Japan also, well-known economist Takashi Negishi (1989) wrote an outstanding book *History of Economic Theory* a la John Hicks. In postscript of the book, Negishi eloquently remarked. "The currently dominating mainstream theory is not the only possible theory. The study of the history of economic theory is important, not only from the heretical point of view, to show how and why the current mainstream theory is wrong, but also from the mainstream point of view, to spur on the development of its own thinking." (Negishi 1989, p. 385.)

popular yet sarcastic essay "*A study of "kū-ki"* (invisible social pressure)," arguing that the heart of ordinary people with cool head sometimes tends to get hot and excited under impersonal and social pressure. For instance, at the end of the Second World War, many young and talented Japanese pilots dared to take on zero fighter planes and carry out "*kamikaze*" (suicidal attack missions with no hope of victory). And now, in parallel with the emergence of the economics of uncertainty, we can see the coming and growing wave of behavioral economics, with Akerlof and Schiller (2009), Kahneman (2003), and Thaler (2015) being recent leaders, in which many biases and irregular human behavior cannot be neglected and should be taken care of in standard social sciences.

In conclusion, we can argue that revealed preference theory a la Samuelson and Houthaker is no longer perfect, having many weaknesses and limitations. Not only that, all theories in economics and related fields are not free of shortcomings and defects. I do declare, however, that Samuelson's revealed preference theory is still alive today and can be further developed. We can still learn new lessons from old teachings.[8]

References

Akerlof GA, Schiller RJ (2009) Animal spirits: how human psychology drives the economy, and why it matters for global capitalism. Princeton University Press

Allen RGD (1936) Professor's Slutsky's theory of consumer's choice. Rev Econ Stud 3:120

Boulding K (1948) Samuelson's foundations of economic analysis: a book review. J Polit Econ 56: 187

Chambers E, Cristopher F (2016) Revealed preference theory. Cambridge University Press

Chipman JS, Hurwicz L, Richter MK, Sonnenschein HF (eds) (1971) Preferences, utility, and demand. Harcourt, New York

Fox KA et al (ed) (1969) Economic models, estimation and risk programming. Springer-Verlag, Berlin

Galbraith JK (1977) The age of uncertainty. Houghton Mifflin, Boston

Hicks, JR (1st ed. 1939, 2nd ed. 1946) Value and capital. Oxford University Press

Hicks JR (1956) A revision of demand theory. Oxford University Press

Hicks JR (1969) A theory of economic theory. Oxford University Press

Houthakker H (1950) Revealed preference and the utility function. Economica 17:159–170

Kahneman D (2003) Maps of bounded rationality: psychology for behavioral economics. Am Econ Rev 93(5):1445–1475

Malinvaud, E (1972, revised ed. 1985) Lectures on microeconomic theory. Elsevier Science Publishers, Amsterdam

McKenzie LW (1957) Demand theory without a utility index. Rev Econ Stud 24:185–189

Negishi T (1989) History of economic theory. North-Holland, Amsterdam

Nicholson W (2005) In: Mason T (ed) Microeconomic theory: basic principles and extensions. Ohio

Sakai Y (1973) Equivalence of the weak and strong axioms of revealed preference without demand continuity assumptions: a "regularity condition" approach. J Econ Theory 8(3):292–304

[8]For the reevaluation of revealed preference in the twenty-first century, see Nicholson (Nicholson 2005), Szenberg et al. (Szenberg et al. 2006), Varian (2006), and Chambers and Cristopher (2016).

Samuelson PA (1938) A note on the pure theory of consumers' behaviour. Economica, New Series 5(17):61–71
Samuelson PA (1947) Foundations of economic analysis. Harvard University Press
Samuelson, PS (1955, 7th ed. 1967) Economics: introductory analysis. McGraw Hill, New York
Sen A (1977) Rational fools: a critique of the behavioural foundations of economic theory. Philos Public Aff 6:317–344
Sen A (1987) On ethics and economics. Blackwell, Oxford
Szenberg M, Ramrattan L, Gottesman AA (eds) (2006) Samuelsonian economics and the twenty-first century. Oxford University Press
Thaler R (2015) Misbehaving: the making of behavioral economics. Norton, New York
Tsuru S (ed) (1961) Has capitalism changed? Iwanami Publishers, Tokyo
U.S.S.R. Academy Economics Institute (Japan edition 1959) Economics textbook. U.S.S.R. Academy, Moscow
Uzawa, H (1960) Preference and rational choice in the theory of consumption. In: Arrow KJ, Karlin S, Suppes P (eds) Mathematical methods in the social sciences. Stanford University Press. p. 129–148. Also contained in Newman P (ed) (1968) Readings in mathematical economics, vol 1, Johns Hopkins Press. p. 297–316
Varian HR (2006) Intermediate microeconomics: a modern approach, 7th edn. East-West Press, New Delhi
Yamamoto S (1983) A study of *kuh-ki* or air ethic (in Japanese). Yamamoto Publishers, Tokyo

Chapter 5
Revealed Favorability and Indirect Utility: A Modern Approach

Abstract This chapter studies the axiomatic foundations of the indirect utility function, based on a revealed preference approach a la Paul A. Samuelson. We look at a chain of comparisons of budgets as if it gives a relation on the normalized-price space (namely, a *revealed favorability relation*) rather than a relation on the commodity space (namely, a *revealed preference relation*). In analogy to the weak and strong axioms of revealed preference, the weak and strong axioms of revealed favorability are newly introduced, and a fundamental theorem concerning the relationship between the latter two axioms is established. Then, the indirect and direct utility functions are effectively derived on the basis of the strong axiom of revealed favorability. It is noted that neither the continuity of the demand function nor the convexity of its range is required for the approach taken here.

Keywords Revealed favorability · Weak and strong axioms · Indirect and direct utility functions · Revealed preference · Paul A. Samuelson

5.1 Revealed Preference and Revealed Favorability: Duality Relations in Consumer Demand

It is Paul A. Samuelson (1947, 1965,1967) who newly adopted a revealed preference approach to consumer choice theory, thus bravely departing from the ordinal utility approach, which had been traditional and dominant for a long time before Samuelson appeared on the economics stage. Such ordinal utility is conveniently shown by a *direct utility function of goods* and has been well developed by E. Slutsky (1915), R. D. D. Allen (1936), John Hicks (1939, 2nd ed. 1946), and others.

This chapter is a completely revised version of Sakai (1977), with many new figures. I am grateful to Lionel W. McKenzie (Rochester), James W. Friedman (Rochester), Hiroshi Atsumi (Toronto), Koji Okuguchi (Tokyo Metropolitan), Marcel K. Richter (Minnesota), Hugo Sonnenschein (Boston), and Werner Hildenbrand (Bonn) for valuable suggestions.

Y. Sakai, *Games, Decisions, and Markets*, New Frontiers in Regional Science: Asian Perspectives 74, https://doi.org/10.1007/978-981-99-5285-4_5

The purpose of this research is to carefully examine the axiomatic foundations of the *indirect utility function of normalized-prices,* which was later developed as a dual of revealed preference approach a la Samuelson (1947, 7th ed. 1967). Remarkably, there has been a strong revival of interest in the indirect utility function since it was first studied by Hotelling (1932) and others a long time ago.

It is now well known that there exist the basic duality relations between the direct and indirect utility functions: namely, maximizing the direct utility of commodity is equivalent to minimizing the indirect utility of prices and income, with the identical budget constraint imposed on both instances. In conjunction with such duality relation, a number of propositions on the structure of utility functions and demand systems have been established by many economic theorists. It is of great interest, therefore, to reexamine the duality relations in the light of revealed preference theory, which have been long neglected so far.

In order to visually understand the duality relationship between the direct and indirect utility approaches, let us consider the simple world with two commodities, x_1 and x_2, and their prices, p_1 and p_2, for illustrative purpose. If we divide those prices by income, then we readily obtain normalized prices q_1 and q_2 in which $q_1 = p_1/m$ and $q_2 = p_2/m$. In the traditional consumer choice world, a representative consumer is supposed to maximize his (direct) utility $U(x_1, x_2)$ subject to his income budget $q_1x_1 + q_2x_2 \leqq 1$. It follows from the resulting consumer equilibrium that x_1 and x_2 are functions of the price pair (q_1, q_2), so that we may duly write $x_1 = x_1(q_1, q_2)$ and $x_2 = x_2(q_1, q_2)$.

If we substitute $x_1(q_1, q_2)$ and $x_2 = x_2(q_1, q_2)$ for the direct utility function $U(x_1, x_2)$, then we can rightly obtain the following indirect utility function:

$$V(q_1, q_2) = U(x_1(q_1, q_2), x_2(q_1, q_2)). \tag{5.1}$$

It is reasonable for us to say that the function V is *indirect utility function* because the utility is *indirectly related to* the normalized-price pair (q_1, q_2) through the "medium" of the commodity pair (x_1, x_2). With the preparations aforementioned, we are ready to say that the consumer is supposed to minimize his indirect utility $V(q_1, q_2)$ subject to the budget set $x_1q_1 + x_2q_2 \leqq 1$.

The dual and symmetrical relationship between the indirect and direct utility function approaches may vividly be visualized in Fig. 5.1. In the upper panel (A), the (normalized) price point q^0 *minimizes the indirect utility* $V(q_1, q_2)$ subject to the budget set $x_1q_1 + x_2q_2 \leqq 1$. In the lower panel (B), the commodity point x^0 *maximizes the direct utility* $U(x_1, x_2)$ subject to the budget set $q_1x_1 + q_2x_2 \leqq 1$.

We are now in a position to exactly define the normalized-price analog of revealed preference relation as follows. First, let us first suppose that a commodity bundle x^1 is chosen from the budget set $b(q^1)$ associated with a normalized price vector q^1. Second, let us also assume that this bundle x^1 belongs to $b(q^0)$, namely the budget set associated with a different normalized-price vector q^0. Then, we duly say that *the new budget set* $b(q^0)$ *is directly revealed more favorable than the old budget set* $b(q^1)$. Or more simply, we can say that *the normalized-price vector* q^0 *is directly revealed more favorable than the normalized-price vector* q^1. This is presumably

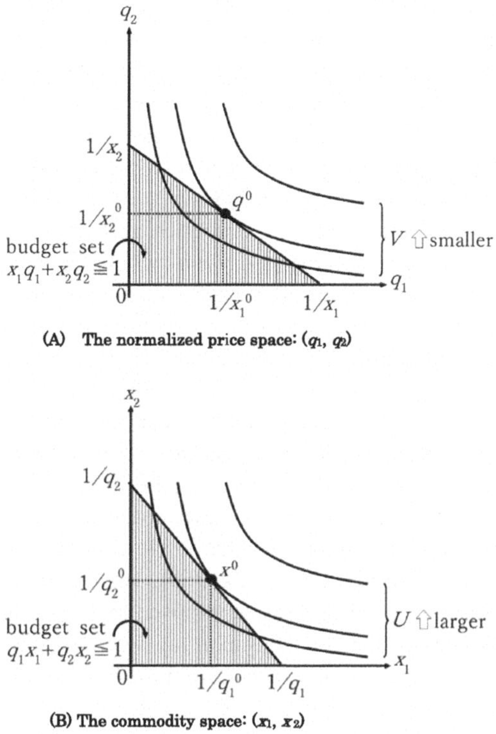

Fig. 5.1 The dual relationship between the indirect and direct utility approaches. (**A**) The normalized price space: (q_1, q_2). (**B**) The commodity space: (x_1, x_2)

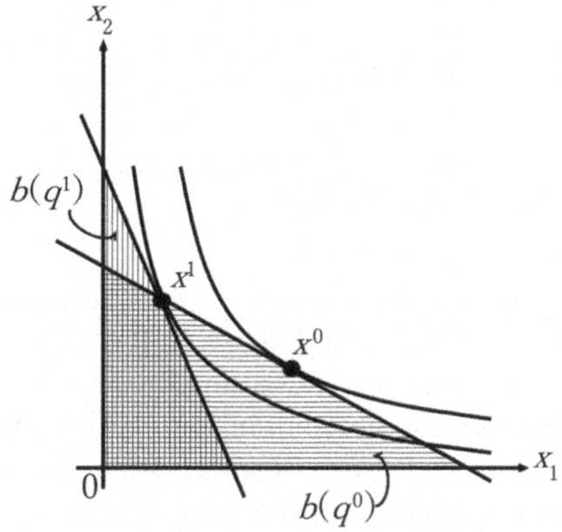

Fig. 5.2 Revealed favorability relation. $b(q^0)$ is directly revealed more favorable than $b(q^1)$; or simply, q^0 is directly revealed more favorable than q^1

because there is some commodity bundle, say x^0, in the new budget set b (q^0), which is better than all commodity bundles in the old budget set b (q^1). Such *revealed favorability relation* may be well illustrated in Fig. 5.2.[1]

[1]The term "more favorable" was first used by Weddepohl (1970).

In the traditional revealed preference theory a la Samuelson (1947) and Houthakker (1950), there exist the famous two axioms. They are the weak and strong axioms of revealed preference. In a very similar way, we can rightly define the normalized-price analogs of those axioms. More specifically, *the weak axiom of revealed favorability* requires that the direct revealed favorability relation thus defined be asymmetric. Needless to say, this is nothing but the normalized-price counterpart of revealed preference on the commodity space. Likewise, analogous to the indirect revealed preference relation on the commodity space, the *indirect revealed favorability relation* may be defined as what we can call "transitive closure" of the direct revealed favorability relation on the price-income space. *The strong axiom of revealed favorability* requires that the indirect revealed favorability relation be asymmetric.

In the traditional revealed preference approach, it is always assumed that the range of the demand function has the convex property. In case the new revealed favorability approach is adopted, however, it will be seen that such a stringent assumption can be discarded; therefore, the results obtained in this chapter help to clarify the role by the convexity condition on the range of the demand function in the theory of consumer's demand. It should also be noticed that, in contrast to the previous results of Kuga (1969), Weddepohl (1970), and others, the continuity or Lipschitz condition is not imposed at all on the demand function here.[2]

The contents of the remaining sections may be outlined in the following way. In Sect. 5.2, a set of exact definitions and detailed assumptions used throughout this paper is carefully introduced. Section 5.3 thoroughly discusses the question whether there exists the duality relation between revealed favorability relations on the normalized-price space and reveled preference relations on the commodities. In Sect. 5.4, we establish an important theorem concerning the relationship between the weak and strong axioms of revealed favorability. As will be seen below, the strong axiom holds if and only if the weak axiom *plus a certain "regularity condition"* both hold. Interestingly enough, this regularity condition is a relationship between two definitions of *income compensated functions* that must hold if the strong axiom holds.

Section 5.5 deals with the *representation problem* of revealed favorability relations. It is shown that, given the strong axiom of revealed favorability together with some other usual assumptions on the demand function, there exists a real-valued function (namely, an indirect utility function) of normalized-price vectors, with the desired properties of minimality, lower semi-continuity, strict quasi-convexity, and monotonicity. A nice bridge between the indirect and direct utility functions will also be built in this research whenever the inverse of the demand function exists. Final several remarks will be made in Sect. 5.6.

---

[2]For example, see Hurwicz and Richter (1971), Richter (1966, 1971), Uzawa (1960, revised 1971), and many others. Strangely enough, it is rather common to assume that the demand function has the *convex property*. One of main purposes in this paper is to do away with such deep-rooted tradition.

5.2 Definitions and Assumptions

In this section, we will introduce certain definitions and basic assumptions to be used throughout this research. In particular, revealed favorability relations on the set of price-income vectors will rigorously be defined in terms of a system of axioms.

We are concerned with the behavior of a rational consumer, who chooses a set of commodity bundles subject to market prices and incomes. The *commodity space Y* is the set of all conceivable commodity bundles. For convenience, we assume that it is the non-negative quadrant of the n-dimensional vector space R^n:

$$Y = \{y : y = (y_1, \ldots, y_n) \text{ and } 0 \leqq y \in R^n\}, \tag{5.2}$$

where y_i denotes the quantity of commodity i for i = 1, ..., n. It should not be necessary that every $y \in Y$ is chosen subject to some normalized-price configuration.

The *price-income space* $P \times M$ is the positive quadrant of the $(n + 1)$-dimensional vector space:

$$P \times M = \{(p, m):(p, m) = (p_1, \ldots, p_n, m) \text{ and } 0 < (p, m) \in R^{n+1}\}, \tag{5.3}$$

where p_i denotes the price of commodity i (i = 1, ... , n), and m represents the consumer's income.

Henceforth, we will also use another price concept by means of "normalization." To this end, let us define the *normalized-price space Q* as the positive quadrant of the n-dimensional vector space:

$$Q = \{q : q = (q_1, \ldots, q_n) \text{ and } 0 < q \in R^n\}, \tag{5.4}$$

where $q_i = (1/m)/p_i$ for any i = 1, ..., n.

As can easily be seen, the normalized-price space Q is actually the set of all conceivable combinations of "normalized-prices," in the sense that "the sum of all x_i -weighted q_i' s over i" is always unity: namely, $\sum_i x_i q_i = 1$. Because of such normalization process, the income component is safely dropped out of our discussion. It is noted that both the commodity space Y and the normalized- price space Q have the same n-dimension.[3]

For a given price vector $q \in Q$, the *budget set b* (q) is defined as follows.

$$b\,(q) = \{y \in Y : q\,y \leqq 1\}. \tag{5.5}$$

Let B be the family of all budget sets:

[3] It is noted that in the present case, Ω is nothing but the positive orthant of R^n and hence happens to be equal to P.

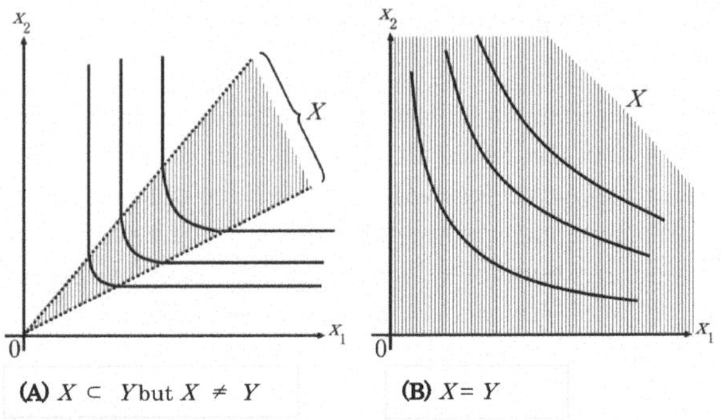

Fig. 5.3 The relationship between X and Y

$$B = \cup \{b\ (q) : q \in Q\}. \tag{5.6}$$

Now, let h be a non-empty *demand correspondence* (*function*) on B, that is, a function that to each $b\ (q)$ assigns a non-empty subset. Such a subset may be called a *choice set,* being written as $h(b(q))$. In what follows, we will obey the following conventions. That is to say, for $h(b(q))$ and $b(q) \in B$, we will also simply write $h(q)$ and $q \in B$, respectively. Further, in the light of (5.5), we let X be the *range of the demand correspondence* h:

$$\begin{aligned} X &= b(B) = \cup \{b(q) : q \in Q\} \\ &= \{x \in Y : x \in h(q) \text{ for some } q \in Q\}. \end{aligned} \tag{5.7}$$

In general, we find $X \subset Y$, but X need not be identical to Y; indeed, X may be a proper subset of Y. This point may more clearly be seen by comparing the two figs. (A) and (B) in Fig. 5.3.

Besides, we make the following usual assumption:

(***H***) For all $q \in Q$ and all $x \in h\ (q)$, we have $qx = 1$.

The assumption (***H***) means that h is (positively) homogeneous of degree zero with respect to q and that the whole budget is spent.

Now, we are in a position to exactly define *revealed favorability relation* on Q in terms of h and b as follows. Suppose that the budget set $b\ (q^0)$ associated with a price vector q^0 contains the choice set $h\ (q^1)$ associated with a *distinct* price vector q^1. Then, we say that $b\ (q^0)$ is *directly revealed more favorable* than $b\ (q^1)$. For this relation, we will simply write $b\ (q^0)F^1\ b\ (q^1)$, since there is presumably some commodity bundle in $b\ (q^0)$, which is better than all commodity bundles in $b\ (q^1)$. We can formally write such relationship as follows.

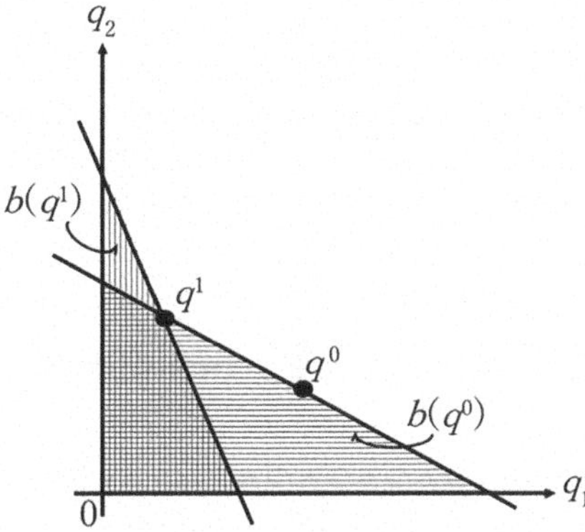

Fig. 5.4 Revealed favorability relation on the *normalized-price space*: $q^0F^1q^1 \Leftrightarrow b\,(q^0)\,F^1\,b\,(q^1) \Leftrightarrow b\,(q^0) \supset h\,(q^1)\ \&\ q^0 \neq q^1$

$$\begin{aligned} &b(q^0)F^1b(q^1) \\ &\Leftrightarrow b(q^0) \supset h(q^1) \text{ and } q^0 \neq q^1 \\ &\Leftrightarrow q^0x^0 = 1 \geqq q^0x^1 \text{for } x^0 \in h(q^0), x^1 \in h(q^1), \text{and } q^0 \neq q^1. \end{aligned} \tag{5.8}$$

In order to avoid the lengthy expression $b\,(q^0)F^1b\,(q^1)$, it is more convenient for us to simply write $q^0F\,^1q^1$. We believe that Fig. 5.4 is helpful for understanding the true meaning of Eq. (5.8). It is noted that although the essence of revealed favorability relation $q^0F^1q^1$ was already pointed out in the last Fig. 5.2, it was quite unfortunate that it was shown on the *commodity space* rather than the *normalized-price* space. To correct such a mismatch, we can draw the new Fig. 5.4, in which the revealed favorability relation $q^0F^1q^1$ is now rightfully on the corresponding normalized-price space. For this point, see Kuga (1969), Weddepohl (1970), and Sakai (1977).

Now, suppose that for a finite sequence r^1, ..., $r^k \in Q$, we find the following sequence of relations:

$$q^0\ F^1\ r^1F^1 \ldots F^1\ r^k\ F^1\ q^1. \tag{5.9}$$

Then, we can say that q^0 is *revealed more favorable in k steps* than q^1, and we write $q^0\ F^k\ q^1$. Moreover, if there exists a certain finite integer k for which the relation $q^0\ F^k\ q^1$ holds, then we should say that q^0 is *indirectly revealed more favorable* than q^1, and we simply write $q^0F\ q^1$. In other words, F is the "transitive closure" of F^1 or the smallest transitive relation including F^1 on Q. It is easy to see that $q^0 \leq q^1$ implies $q^0F^1\ q^1$ and hence $q^0\ F\ q^1$.

Corresponding to these two relations F^1 and F on Q, let us introduce the two axioms of revealed favorability, namely, the *weak axiom* **(*WF*)** *and the strong axiom* **(*SF*)** in the following fashion:

(*WF*) For q^0, $q^1 \in Q$, q^0 F^1 q^1 implies $\sim q^1F^1q^0$,

(*SF*) For q^0, $q^1 \in Q$, $q^0\ F\ q^1$ implies $\sim q^1\ F\ q^0$,where, in general, the expression " $\sim$ *ABC* " stands for the "negation of *ABC*." These two axioms assert that the consumer's behavior should be "directly" or "indirectly" consistent; therefore, they are quite analogous to the traditional revealed preference axioms of Samuelson (1947) and Houthakker (1950) on *Y*, the commodity space. The duality relationship between revealed favorability relations on *Q* and revealed preference relations on *Y* will be investigated in more details in the next section.

5.3 Revealed Favorability Versus Revealed Preference

In this section, we will examine whether there exists the duality relationship between revealed favorability on the price space and revealed preference on the commodity space. We are also concerned with the question of how this duality relationship corresponds to the more familiar one between the indirect and direct utility functions.

As mentioned above, revealed preference relations on *Y* are defined in terms of *h* and *b* as follows. If there is $q \in Q$ such that $x \in h(q)$, $y \in b(q)$, and $x \neq y$, then we say that *x* is *directly revealed preferred* to *y*, and we write $x\, S\, y$. If there is some finite sequence $u^1, ..., u^k$ of elements of *Y* such that $x\, S\, u^1\, S \ldots S\, u^k\, S\, y$, then we say that *x* is *indirectly revealed preferred* to *y*, and we write $x\, H\, y$.

On the one hand, the *weak axiom* **(*WP*)** *of revealed preference* requires that the direct revealed preference relation a la Samuelson *S* be non-symmetric:

(*WP*) For $x, y \in Y$, $x\, S\, y$ implies $\sim y\, S\, x$.

On the other hand, the *strong axiom* **(*SP*)** *of revealed preference* requires that the indirect revealed preference relation *H be* non-symmetric:

(*SP*) For $x, y \in Y$, $x\, H\, y$ implies $\sim y\, H\, x$.

It is easy to show that if *h* satisfies **(*WF*)** on *Q*, then for any $x \in X$, the inverse image of *x* by *h* contains a single element; therefore, *h* is uniquely invertible. In contrast, it is also easily seen that if *h* satisfies **(*WP*)** on *X*, then for any $q \in Q$, the image of *q* by *h* is a singleton; therefore, *h* is single-valued. Hence, **(*WF*)** does not imply **(*WP*),** and vice versa. It is equally clear that **(*SF*)** does not imply **(*WF*)**, and vice versa. This point will clearly be understood by the help of graphical illustrations.

On the one hand, let us consider simple examples of indifference curves on *Q* with "pointed" portions, or equivalently, those of indifference curves on *X* with "flat" portions. As is seen in Fig. 5.5(A) and (B), they indicate the consumer satisfying **(*WF*),** but not **(*WP*)**. On the other hand, as is clear in Fig. 5.6(A) and (B), symmetrical examples of indifference curves on *Q* with "flat" portions (and hence those of indifference curves on *X* with "pointed" portions) indicate the consumer satisfying, **(*WP*)** but not **(*WF*)**. If *h* happens to be a one-to-one correspondence between *Q* and *X*, however, then it is readily seen that **(*WF*)**holds on *Q* if and only if **(*WP*)** holds on *X,* and also that (SP) holds on *Q* if and only if **(*SP*)** holds on *X*.[4]

[4]I am indebted to Richter (1966) for clarifying this line of argument.

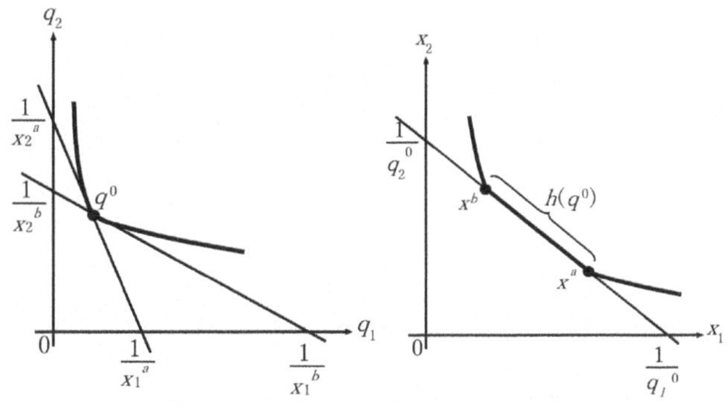

(A) Indifference curves have "pointed" sections on *X*.

(B) Indifference curves have "flat" sections on *Q*.

Fig. 5.5 The consumer satisfies **(*WF*)**, but not **(*WP*)**

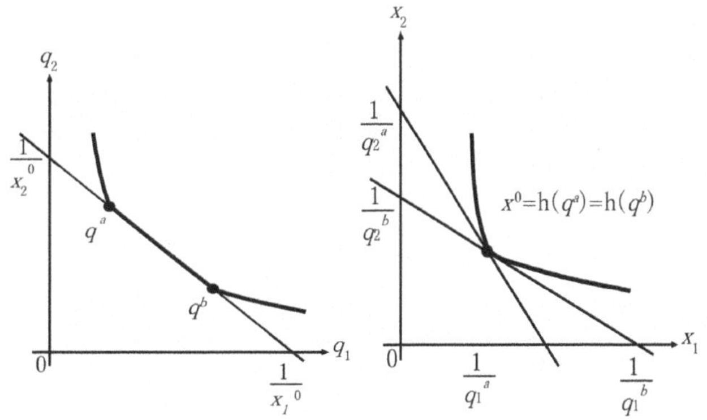

(A) Indifference curves have "flat" sections on *Q*.

(B) Indifference curves have "pointed" sections on *X*.

Fig. 5.6 The consumer satisfies **(*WP*)**, but not **(*WF*)**

5.4 Relationship Between the Weak and Strong Axioms of Revealed Favorability

In this section, we will study the relationship between the weak and strong axioms of revealed favorability. Let us suppose that the demand function h satisfies the homogeneity assumption **(*H*)**. Then, it will be seen that the strong axiom **(*SF*)** of revealed favorability holds if and only if the weak axiom **(*WF*)** of revealed favorability together with a certain "regularity condition" holds. As will be seen, such a

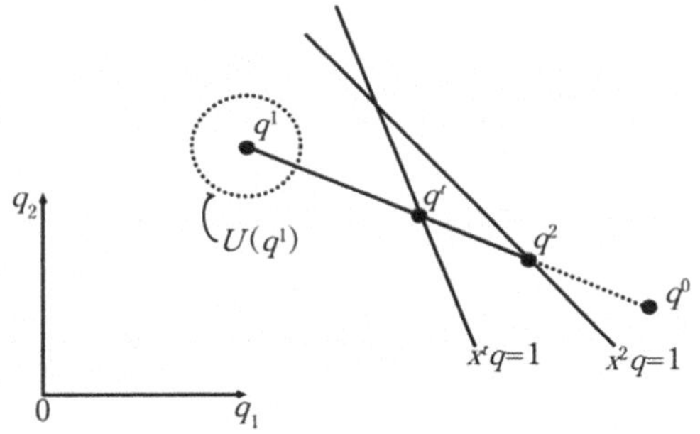

Fig. 5.7 The proof of Lemma 5.1 is well illustrated here

regularity condition is a relationship between two kinds of "income compensated functions " that must hold if the strong axiom holds.

To this end, we first establish the following important lemma.

Lemma 5.1 (The Closeness of the Set $\{q \in Q: \sim q\, F\, q^0\}$) Suppose that the demand function h satisfies **(*H*)** and **(*WF*)**. Then, for any $q \in Q$, the set $\{q \in Q: \sim q\, F\, q^0\}$ is closed in Q.

Proof[5] It suffices to show that the set $\{q \in Q: q\, F\, q^0\}$ is open in Q. Let us suppose $q^1 F\, q^0$. Then, by the definition of F, there exists a certain sequence $q^1, q^2, ..., q^0$ such that $q^1F^1q^2F^1 ... F^1\, q^0$. In the light of Eq. (5.8), this surely implies that there exists a price vector $q^2 \in Q$ such that

$$q^1\, x^1 = 1 \geqq q^1\, x^2 \text{ for } x^1 \in h\left(q^1\right), x^2 \in h\left(q^2\right), \text{and } q^1 \neq q^2. \tag{5.10}$$

and

$$q^2\, F\, q^0 \text{ or } q^2 = q^0. \tag{5.11}$$

Now, as is seen in Fig. 5.7, let us define $q^t = (1 - t)q^1 + t{\cdot}q^2$, $t \in (0, 1)$, and $x^t \in h\,(q^t)$. Then, we must have the following equations.

$$q^t\, x^2 = (1-t)q^1x^2 + t \cdot q^2\, x^2 = (1-t)q^1x^2 + t. \tag{5.12}$$

We easily note the following equations.

$$q^t\, x^t = 1 \geqq q^t\, x^2, q^t \neq q^2. \tag{5.13}$$

In the light of Eq. (5.12) and the weak axiom **(*WF*)**, we must obtain the following.

[5] I am thankful to Uzawa (1960) for suggesting the proof presented here.

$$q^2 x^t > 1 = q^2 x^2. \tag{5.14}$$

Now, we recall that by the choice of q^t, we must find the following equation.

$$\begin{aligned}&(1-t)(q^1x^t-1)+t(q^2x^t-1)\\&=[(1-t)q^1+t\cdot q^2]x^t+[(1-t)(-1)+t(-1)]\\&=q^tx^t-1=0.\end{aligned} \tag{5.15}$$

If we compare Eqs. (5.12) and (5.13), we clearly obtain $q^1x^t < 1$. We can choose a neighborhood $U\,(q^1)$ of q^1 so that the following equation holds.

$$q^1x^t < 1 = q\,x, x \in h\,(q) \text{ for any } q \in U\left(q^1\right). \tag{5.16}$$

Equations (5.14), (5.11), and (5.10) taken together yield the following relation: q^2 $F\,q^0$ for any $q \in U\,(q^1)$. This ensures that the set $\{q \in Q\text{: } q\;F\;q^0\}$ is open in Q.
Q.E.D.

As our experience often teaches us, a graphical illustration may be a great help for our complicated mathematical proof. So, we believe that Fig. 5.7 may well illustrate the essence of the proof of Lemma 5.1.

Now, for a given $p \in Q$ and a given $q^0 \in Q$, let us define the "M-plus and M-minus sets" of incomes as follows:

$$\boldsymbol{M}+(\boldsymbol{p},\boldsymbol{q}^{\mathbf{0}})=\{m\in M:\sim(1/m)\,p\;F\;q^0\}, \tag{5.17}$$

$$\boldsymbol{M}-(\boldsymbol{p},\boldsymbol{q}^{\mathbf{0}})=\{m\in M:\sim q^0\;F\;(1/m)\,p\}. \tag{5.18}$$

While the "M-plus set" or $\boldsymbol{M+}$ $(\boldsymbol{p},\boldsymbol{q}^{\mathbf{0}})$ represents the set of incomes m such that (1/m)p is *not* indirectly revealed more favorable than q^0, the "M-minus set" or $\boldsymbol{M}-$ $(\boldsymbol{p},\boldsymbol{q}^{\mathbf{0}})$ represents the set of incomes m such that q^0 is *not* indirectly revealed more favorable than (1/m)p . Then, the "m-plus and m-minus income compensation functions," ***m+*** and ***m-***, can respectively be defined in terms of ***M***+ and ***M***− in the following way[6]:

$$\boldsymbol{m}+\left(\boldsymbol{p},\boldsymbol{q}^{\mathbf{0}}\right)=\boldsymbol{sup}\left\{m:m\in\boldsymbol{M}+\left(\boldsymbol{p},\boldsymbol{q}^{\mathbf{0}}\right)\right\}, \tag{5.19}$$

$$\boldsymbol{m}-\left(\boldsymbol{p},\boldsymbol{q}^{\mathbf{0}}\right)=\mathbf{inf}\left\{m:m\in\boldsymbol{M}-\left(\boldsymbol{p},\boldsymbol{q}^{\mathbf{0}}\right)\right\}. \tag{5.20}$$

We can give economic interpretations to those two functions ***m+*** and ***m-*** in the following fashion. Let p and q^0 respectively be a given "absolute" price vector and a given "relative" price vector q^0. Then, whereas the function $\boldsymbol{m}+(\boldsymbol{p},\boldsymbol{q}^{\mathbf{0}})$ stands for the

[6]Income compensation functions were first introduced by McKenzie (1957) and Yokoyama (1953), independently, in terms of preference orderings on the commodity space.

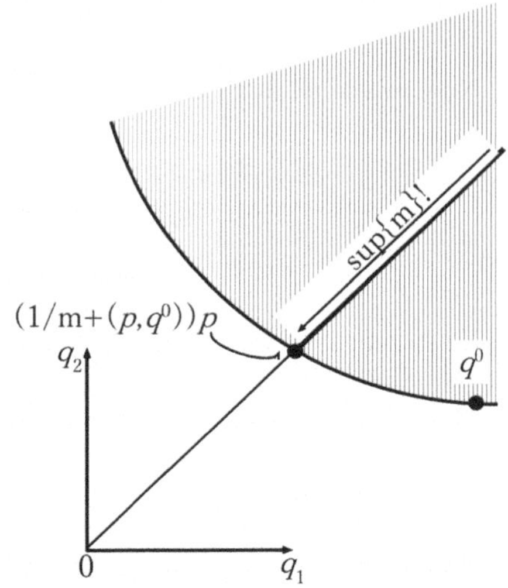

Fig. 5.8 The "m-plus income compensation function": $m + (p, q^0)$

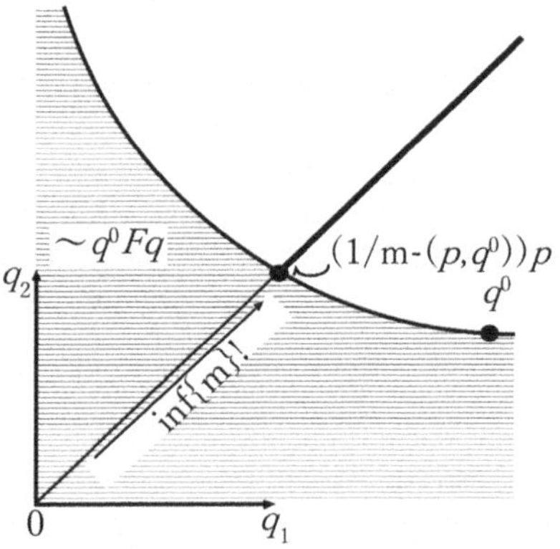

Fig. 5.9 The "m-minus income compensation function": $m - (p, q^0)$

supremum (or the least upper bound) of incomes m such that $(1/m)\,p$ is *not* indirectly revealed more favorable than q^0, the function $\boldsymbol{m - (p, q^0)}$ stands for the *infimum* (or the greatest lower bound) of incomes m such that q^0 is *not* indirectly revealed more favorable than $(1/m)\,p$. It is clear by definition that $\boldsymbol{M + (p, q^0)} \supset (0, \boldsymbol{m + (p, q^0)})$ and $\boldsymbol{M - (p, q^0)} \supset (\boldsymbol{m - (p, q^0)}, +\infty)$.

While the "m-plus income compensation function" $\boldsymbol{m + (p, q^0)}$ corresponding to the "M-plus set" $\boldsymbol{M + (p, q^0)}$ is illustrated in Fig. 5.8, the "m-minus income compensation function" $\boldsymbol{m - (p, q^0)}$ corresponding to the "M-minus set" $\boldsymbol{M - (p, q^0)}$ is shown in Fig. 5.9.

Then, we are ready to establish the following useful lemma.

Lemma 5.2 (The Closeness of the M-plus Set or $\boldsymbol{M + (p, q^0)}$) Let us suppose that the demand function h satisfies the homogeneity assumption (***H***) and the weak axiom (***WF***). Then, for any $(p, q^0) \in P \times Q$, the M-plus set $\boldsymbol{M + (p, q^0)}$ is closed in M.

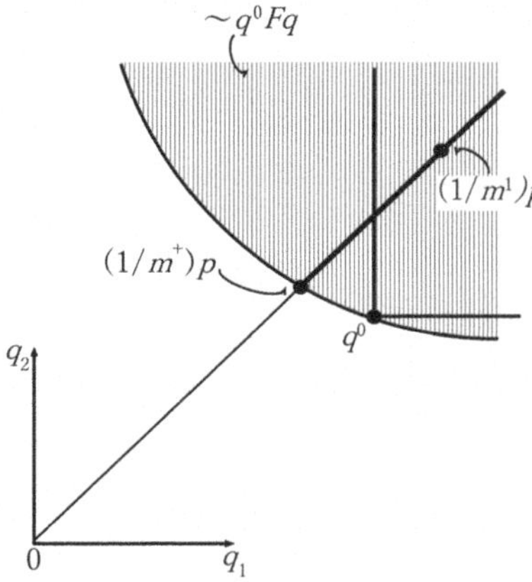

Fig. 5.10 We can choose a *sufficiently small* $m^1 \in M$ such that $(1/m^1)\,p > q^0$

Proof By definition, it is evident that $\boldsymbol{M+(p,q^0)} = M - \{m \in M\text{: } (1/m)\,p\,F\,q^0\}$. So, to prove that the set $\boldsymbol{M+(p,q^0)}$ is closed in M, it suffices to show that its complementary set $\{m \in M\text{: } (1/m)\,p\,F\,q^0\}$ is open in M.

Suppose $(1/m)\,p\,F\,q^0$. Then, in the light of Lemma 5.1, the set $\{q \in Q\text{: } q\,F\,q^0\}$ is open in Q. Therefore, there exists a neighborhood $U((1/m)p)$ of $(1/m)p \in Q$ so that $q^1\,F\,q^0$ for any $q^1 \in U((1/m)p)$.

Choose a sufficiently small $e > 0$ such that $(1/(m+e)), (1/(m-e)) \in U((1/m)\,p)$. Then, for any $m' \in (m-e, m+e)$, we obtain $(1/m')\,F\,q^0$. This indicates that the set $\{m \in M\text{: } (1/m)\,p\,F\,q^0\}$ is open in M.

Q.E.D.

According to Lemma 5,2, we find $\boldsymbol{M+(p,q^0)} = (0, \boldsymbol{m+(p,q^0)}]$ whenever h satisfies **(*H*)** and **(*WF*)**. We are now approaching to the final goal of the equivalence theorem. To reach it safely, we need to pass through the following midpoint.

Lemma 5.3 (The Existence of $\boldsymbol{m+(p,q^0)}$ and $\boldsymbol{m-(p,q^0)}$) Suppose that the demand function h satisfies the homogeneity assumption **(*H*)** and the strong axiom **(*SF*)**. Then, for any $(p, q^0) \in P \times Q$, the income compensated functions $\boldsymbol{m+(p,q^0)}$ and $\boldsymbol{m-(p,q^0)}$ exist and are finite.

Proof To see that $\boldsymbol{m+}$ exists and is finite, it suffices to show that the set $\boldsymbol{M+(p,q^0)}$ is non-empty and bounded above.

Let $(p, q^0) \in P \times Q$. Then, as is seen in Fig. 5.10, we can surely choose a sufficiently *small* $m^1 \in M$ such that $(1/m^1)p > q^0$. Let $x^0 \in h(q^0)$ and $x^1 \in h((1/m^1)\,p)$. Then, we find the following equation.

$$q^0 x^0 = 1 = (1/m^1)p\,x^1 > q^0 x^1, \tag{5.21}$$

so that we must obtain $q^0\,F^1\,(1/m^1)p$, implying $q^0\,F\,(1/m^1)\,p^1$. By virtue of the strong axiom **(*SF*)**, this gives $\sim (1/m^1)\,p^1 F\,p^0$, so that $m^1 \in \boldsymbol{M+(p,q^0)}$. Therefore, $\boldsymbol{M+(p,q^0)}$ is nonempty.

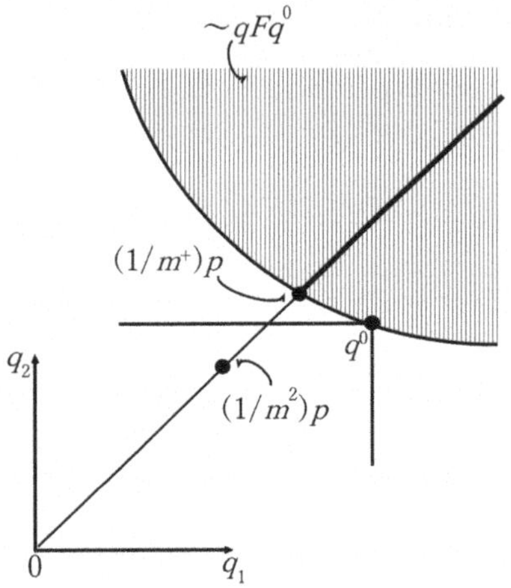

Fig. 5.11 We can choose a *sufficiently large* $m^2 \in M$ such that $(1/m^2)p < q^0$

Now, as is seen in Fig. 5.11, we can choose a sufficiently *large* m^2 such that $m^2 \in M$ such that $(1/m^2)p < q^0$. Letting $x^0 \in h(q^0)$, we have

$$1 = q^0 x^0 > (1/m^2)p\, x^0, \tag{5.22}$$

whence $(1/m^2)\, p\, F\, q^0$. This shows that m^2 is an upper bound to $\boldsymbol{M} + (\boldsymbol{p}, \boldsymbol{q}^0)$.

The proof that $\boldsymbol{m} - (\boldsymbol{p}, \boldsymbol{q}^0)$ exists and is finite proceeds in a similar way.

Q.E.D.

We are ready to formulate a *regularity condition* in terms of income compensation functions $\boldsymbol{m} + (\boldsymbol{p}, \boldsymbol{q}^0)$ and $\boldsymbol{m} - (\boldsymbol{p}, \boldsymbol{q}^0)$:

(***R***) For any $(p, q^0) \in P \times Q$, $\boldsymbol{m} + (\boldsymbol{p}, \boldsymbol{q}^0) \geqq \boldsymbol{m} - (\boldsymbol{p}, \boldsymbol{q}^0)$.

As will be seen, the regularity condition (***R***) plays a critical role in making a bridge between the weak and strong axioms of revealed favorability, (***WF***) and (***SF***). In plain English, it states that for any given price vector (p, q^0), the supremum of incomes m such that $(1/m)\, p$ is "no better than" q^0 is greater than or equal to the infimum of incomes m such that q^0 is "no better than" $(1/m)\, p$. Here again, seeing is really believing! Figure 5.12 represents the case in which (***R***) is satisfied, while Fig. 5.13 indicates the case in which (***R***) is *not* satisfied.

As is quite clear from comparison of these two figures, condition (***R***) requires that there be no "$\sim F$ gaps" in the ray $\{(1/m)\, p: m \in M\}$ for any given price vector p, yet allowing for the existence of "$\sim F$ overlaps" in the relevant ray. It is noted that in these two figures, the upper-shaded area denotes the set $\{q \in Q: \sim q\, F\, q^0\}$, and the lower-shaded area the lower-shaded area represents the set $\{q \in Q: \sim q^0\, F\, q\}$.

Now, it is high time for us to establish a very important theorem concerning the relation the weak and strong axioms of revealed favorability.

Theorem 5.4 (The Main Equivalence Theorem) Suppose that the demand function h satisfies (***H***). Then, the strong axiom (***SF***) of revealed favorability holds if and

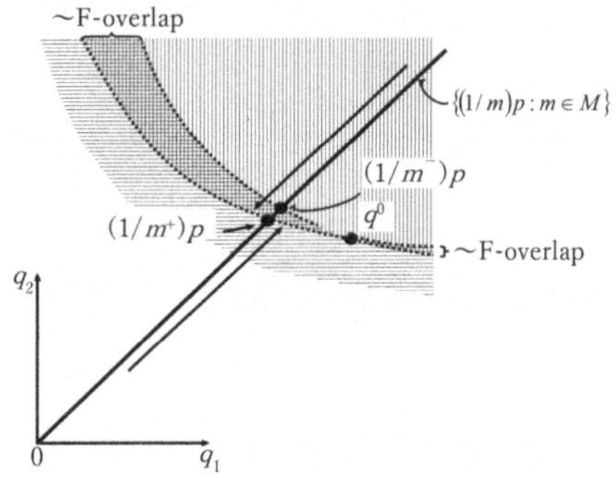

Fig. 5.12 The regularity condition (R) is satisfied: $m+(p, q^0) \geqq m-(p, q^0)$

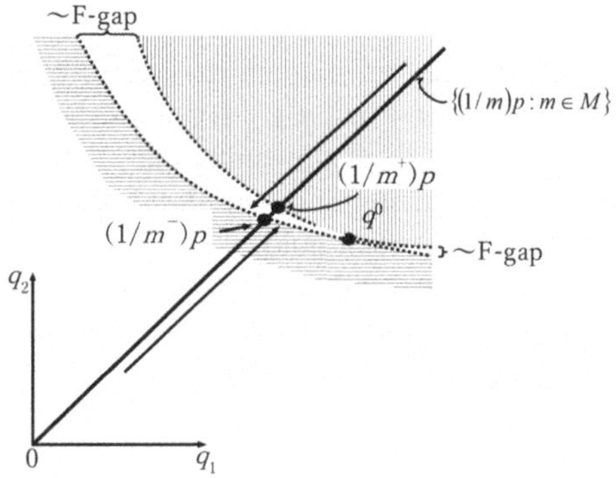

Fig. 5.13 The regularity condition (R) is *not* satisfied: $m+(p, q^0) < m-(p, q^0)$

only if both the weak axiom **(*WF*)** of revealed favorability and the regularity condition **(*R*)** hold.

Proof (a) (Necessity) Suppose that **(*SF*)** holds. Then, **(*WF*)** is obviously implied. Besides, by virtue of Lemma 5.3, the two income compensation functions $\boldsymbol{m+(p, q^0)}$ and $\boldsymbol{m-(p, q^0)}$ surely exist and are finite.

Now, assume by way of contradiction that the regularity condition **(*R*)** does not hold, so we should find $\boldsymbol{m+(p, q^0)} < \boldsymbol{m-(p, q^0)}$ for some $(p, q^0) \in P \times Q$. Choose a "middle point" $m \in M$ so that the following inequalities hold:

$$\boldsymbol{m}+(\boldsymbol{p}, \boldsymbol{q}^{\mathbf{0}}) < m < \boldsymbol{m}-(\boldsymbol{p}, \boldsymbol{q}^{\mathbf{0}}). \tag{5.23}$$

By the definitions of $\boldsymbol{m+(p, q^0)}$ and $\boldsymbol{m-(p, q^0)}$, Eq. (5.23) implies that (1/m) p F q^0 and q^0 F (1/m)p. This clearly contradicts **(*SF*)**. So, to get rid of a contradiction, we must conclude that the regularity condition **(*RF*)** must hold.

(b) (Sufficiency) Suppose that **(*WF*)** and **(*R*)** both hold. Let $q^1, q^0 \in Q$ be such that q^1 F q^0. Then, we will show the following relation.

$$\boldsymbol{m}+(\boldsymbol{q}^{\mathbf{1}}, \boldsymbol{q}^{\mathbf{0}}) < 1. \tag{5.24}$$

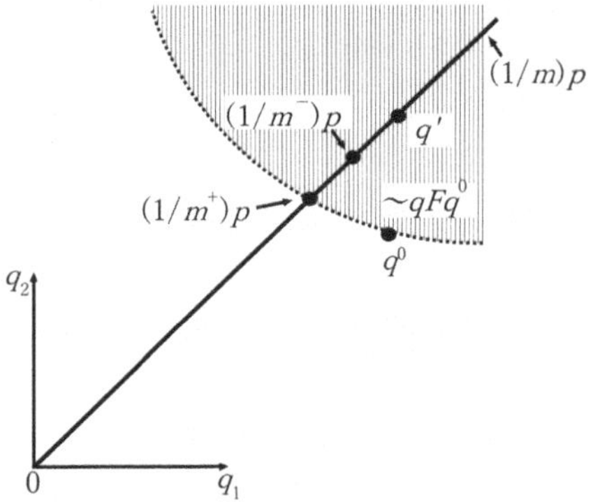

Fig. 5.14 The proof outline of Theorem 5.14 (the sufficiency part): $q^1\ Fq^0$ implies $\sim q^0\ Fq^1$, ensuring (*DF*)

To this end, let us dare to assume otherwise: namely, $\boldsymbol{m} + (\boldsymbol{q}^1, \boldsymbol{q}^0) \geqq 1$. Since **(*WF*)** yields $\boldsymbol{M} + (\boldsymbol{q}^1, \boldsymbol{q}^0) = (0, \boldsymbol{m} + (\boldsymbol{q}^1, \boldsymbol{q}^0)]$by means of Lemma 5.2, it would follow that $1 \in \boldsymbol{M} + (\boldsymbol{q}^1, \boldsymbol{q}^0)$, meaning that $\sim (1/1)\ q^1\ F\ q^0$, or simply $\sim q^1\ F\ q^0$. Clearly, this contradicts our initial assumption. To get rid of a contradiction, we must conclude that Eq. (5.24) must hold.

Now, let us recall the regularity condition **(*R*)**, which says that $\boldsymbol{m} + (\boldsymbol{p}, \boldsymbol{q}^0) \geqq \boldsymbol{m} - (\boldsymbol{p}, \boldsymbol{q}^0)$. If we combine this inequality and Eq. (5.23), then we immediately find the following relation.

$$\boldsymbol{m} - (\boldsymbol{q}^1, \boldsymbol{q}^0) < 1. \tag{5.25}$$

Let us recall that $\boldsymbol{M} - (\boldsymbol{q}^1, \boldsymbol{q}^0) \supset (\boldsymbol{m} - (\boldsymbol{q}^1, \boldsymbol{q}^0), +\infty)$. Then, this clearly implies $1 \in \boldsymbol{M} - (\boldsymbol{q}^1, \boldsymbol{q}^0)$, meaning that $\sim q^0\ F\ (1/1)\ q^1$, or simply $\sim q^0 F q^1$. For the outline of proof here, see Fig. 5.14.

To conclude, we have thus seen that under **(*WF*)** and **(*R*)**, the revealed favorability relation F is non-symmetric, meaning that $q^1\ Fq^0$ implies $\sim q^0\ Fq^1$.

The proof is now complete.

Q.E.D.

As far as we know, the relation between the weak and strong axioms of *revealed favorability* on the normalized-price space has hardly been investigated in the economics literature. Needless to say, this is exactly the dual to the more popular relation between the weak and strong axioms of *revealed preference* on the commodity space, which has been so intensively explored in the literature. Such non-symmetric treatment of the two approaches—one almost neglected, and another extensively explored—seems to be very strange from common sense.[7]

We should point it out that the hypothesis of Theorem 5.4 is not so strong and indeed amazingly rather weak. This is because (1) *no* continuity condition is

[7] Even in the twenty-first century, the relationship between the weak and strong axioms of *revealed favorability on the normalized-price space has rarely been* mentioned in the microeconomics literature. Here again, we see extreme difficulty to break through the hard crust of convention.

Fig. 5.15 h satisfies (SF) and is continuous. However, its range is *not* convex; $x^t = (1\ t)\, x^0 + t\, x^1$ does *not* belong to X. (**A**) The indirect preferences in the (q^1, q^2) space. (**B**) The direct preferences in the (x^1, x^2) space

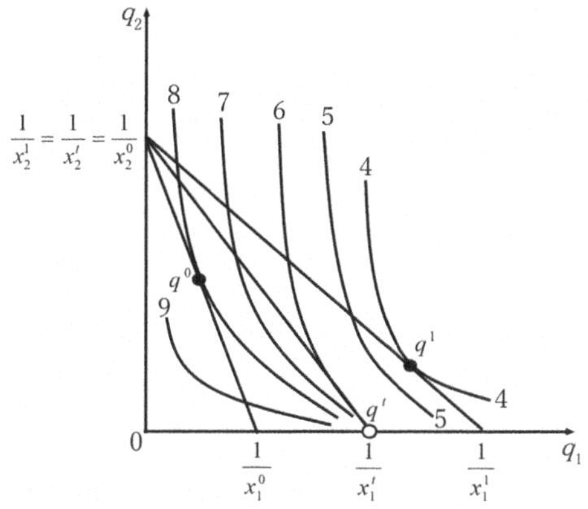

(A) The indirect preferences in the (q^1, q^2) space.

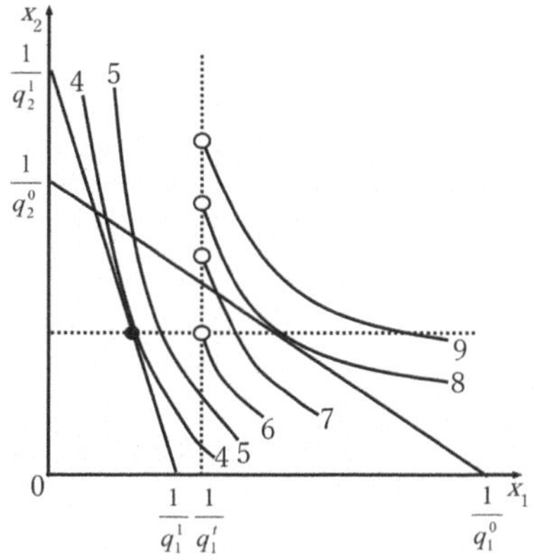

(B) The direct preferences in the (x^1, x^2) space.

imposed on the demand function h, and (2) X, the range of h, need NOT be convex in Y, the whole commodity space. In particular, the above conditions (1) and (2) are *not* implied by the budget assumption ***(H)*** and the strong axiom ***(SF)***. This important point can be confirmed by the two examples shown in Figs. 5.15 and 5.16.

In Fig. 5.15, let h be generated by *indirect* preferences indicated by the upper diagram (A), or equivalently, by *direct* preferences indicated by the lower diagram (B). Then, as can easily seen, h satisfies ***(SF)*** and is continuous. However, the range of h is *not* convex; indeed, we find the middle point $x^t = (1 - t)x^0 + t\, x^1$ does *not* belong the set X.

Fig. 5.16 h satisfies (*SF*), but it is NOT continuous. The range of h is *not* convex either; in fact, $x^t = (1 - t)\, x^0 + t\, x^1$ does *not* belong to X. (**A**) The indirect preferences in the (q^1, q^2) space. (**B**) The direct preferences in the (x^1, x^2) space

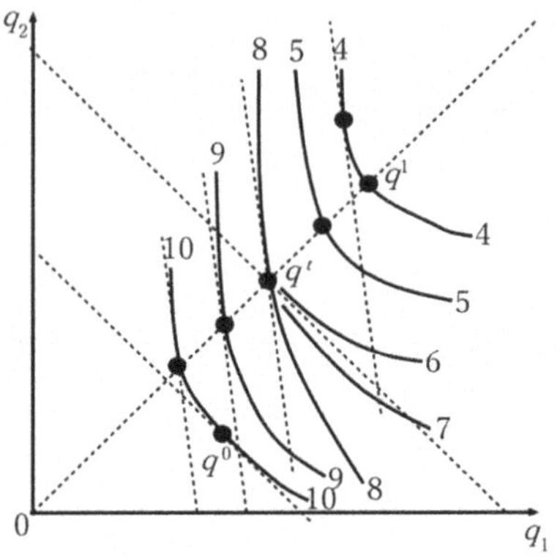

(A) The indirect preferences in the (q^1, q^2) space.

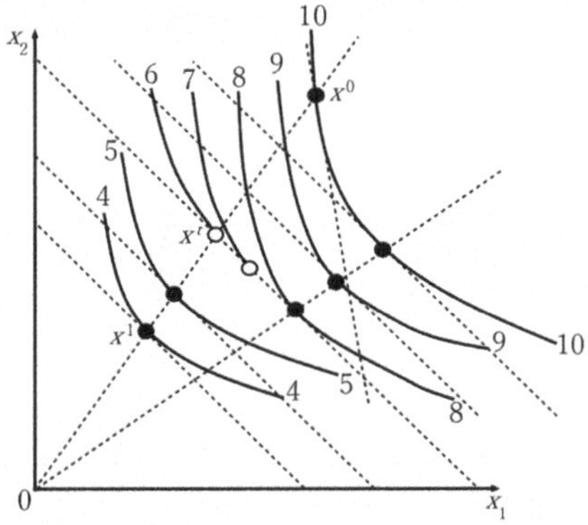

(B) The direct preferences in the (x^1, x^2) space.

In Fig. 5.16, similar yet different diagrams are drawn in (A) and (B). Here again, h satisfies **(*SF*)**, but it is *not* continuous, and its range is *not* convex; indeed, we find that the middle point $x^t = (1 - t)\, x^0 + t\, x^1$ does *not* belong to the set X.[8]

We must keep in mind that situations like Figs. 5.15 and 5.16 are less than extraordinary than they seem, but might happen in many ordinary circumstances.

[8] I am thankful to Hugo Sonnenschein for suggesting Figs. 5.15 and 5.16. In my opinion, he is a very skillful drawer of indifference figures.

5.5 Derivation of the Indirect and Direct Utility Functions

The duality relation between the *direct* and *indirect* utility functions has especially been studied in connection with special functional forms of *separability*. Although such a separability approach is of some mathematical interest, its deeper and wider significance in economic theory remains to be still debatable.[9]

In this section, we would like to investigate more fundamental questions than the separability question aforementioned. First, we will discuss the problem whether the indirect revealed favorability relation on the normalized space can be represented by a real-valued function, i.e., the *indirect* utility function. Second, we will explore the related problem whether there is a way of making a bridge between the *indirect* and *direct* utility functions from the perspective of the revealed favorability approach taken in the present research.

In this connection, we will state and prove the following result of very rich substance.

Theorem 5.5 (The Derivation of the *Indirect* Utility) Suppose that the demand function h satisfies the budget assumption ***(H)*** and the strong axiom of revealed favorability ***(SF)***. Then, there exists a real-valued function v on Q, namely the *indirect utility function*, such that the following series of properties hold.

1. (minimality) For any $q \in Q$, the following equation holds.
   $h(q) = \{x \in X\colon q\, x \leqq 1, \text{ and } (r) \geqq v(q) \text{ for any } r \in Q \text{ such that } r\, x \leqq 1\}.$
2. (closeness) For any $q \in Q$, the set $\{r \in Q\colon v(q) \geqq v(r)\}$ is closed in Q.
3. (strict concavity) If $v(q^1) \geqq v(q^0)$, $q^1, q^0 \in Q$, $q^1 \neq q^0$, and
   $q^t = (1-t)q^1 + t\, q^0$, then $v(q^1) > v(q^t)$.
4. (monotonicity) If $q^1 \leq q^0$ and $q^1, q^0 \in Q$, then $v(q^1) > v(q^0)$.
5. (heritability) For any $q^1, q^0 \in Q$, $q^1 F\, q^0$ implies $v(q^1) > v(q^0)$.

Proof The proof of this theorem is rather long and occasionally mathematically advanced. So, we carefully carry it out in a step-by-step fashion.[10]

(Step 1) Homomorphism:

Since F is a non-symmetric and transitive relation on Q by means of the strong axiom ***(SF)*** of revealed favorability, and definition itself, it should be non-reflexive and transitive on Q.

Let C denote the set of all elements of Q having all coordinates *rational.* Then, we will show that this C is a countable partially F-dense subset of Q. To this end, let us assume that $q^1, q^0 \in Q - C$ satisfy $q^1\ F\ q^0$. Then, by virtue of Lemma 5.1, we can choose a neighborhood $U(q^1)$ of q^1 so that $s\ F\ q^0$ for all $s \in U(q^0)$. Take $c \in U(q^1) \cap C$ for which $c > q^1$. We then have $q^1 F\ c\ F\ q^0$; therefore, C is F-dense in Q. Now, we pay special attention to the following important theorem:

---

[9]For the duality results, see Lau (1969). Also see Diewert (1982).

[10]I am grateful to Richter (1966, 1971) and Hurwicz and Richter (1971) for giving a hint for the proof presented here.

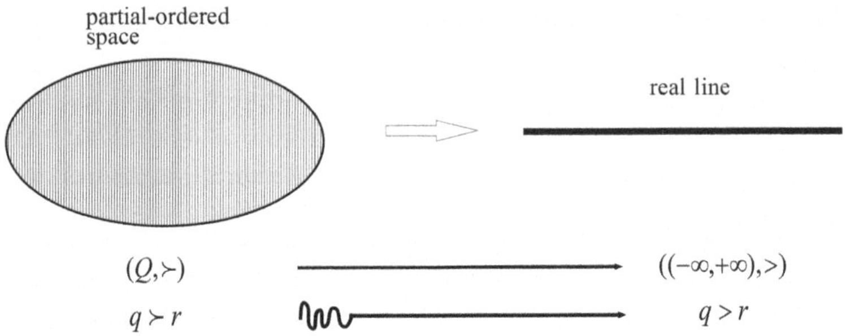

Fig. 5.17 Homomorphism from $(X, \succ)$ into $((-\infty, +\infty), >)$

Theorem (Richter 1971)

Let $\succ$ *be a non-reflective and transitive relation on a set* Q. *If there is a countable partially* $\succ$ *dense subset of* Q, *then there exists a weak homomorphism from the partially ordered space* $(Q, \succ)$ *into the real straight line* $((-\infty, +\infty), >)$.

The proof of this theorem can be seen in Richter (1971), p. 49, and omitted here. Figure 5.17 may give us a rough idea of homomorphism from the partially ordered space $(Q, \succ)$ into the real straight line $((-\infty, +\infty), >)$. If we apply the Richter theorem to the present case, then we can see that there exists a non-negative-valued and bounded function f: $Q \rightarrow (-\infty, +\infty)$, satisfying the following relation.

$$\text{For any } q^1, q^0 \in Q, q^1\ F\ q^0 \Rightarrow f\ (q^1) > f\ (q^0). \tag{5.26}$$

(Step 2) Property (2):

Since the function f introduced above is bounded, we can newly define a real-valued function v on Q by $v = lim\ inf f$: namely, for $q \in Q$, the following equation holds.

$$v\ (q) = sup\ \{ inf\ \{ f\ (s) : s \in U\} : q \in U \in \text{Ц}\}, \tag{5.27}$$

where Ц denotes the family of open sets in Q. Because v is the lower limit function of f, it must be lower semi-continuous. For this point, see McShane & Botts (1952), Theorem 3.6, p. 7.5. Therefore, we can see that the set $\{r \in Q\text{: } v(q) \geqq v\ (r)\}$ is closed in Q, successfully establishing Property (2).

(Step 3) Property (5):

To see Property (5), let us choose q^1, $q^0 \in Q$ such that q^1Fq^0. Since F is lower, lower semi-continuous by Lemma 5.1 above, we can easily choose $q^2 \in Q$ for which $q^1Fq^2\ Fq^0$, as was shown in (Step 1) above. By the lower semi-continuity of F again, q^1Fq^2 implies that there exists a neighborhood $U\ (q^1)$ of q^1 so that $q\ F\ q^2$ for all $q \in U(q^1)$. Consequently, taking advantage the Richter theorem again, we find from (5.25) that the following inequalities hold:

$$inf\ \{\ f(q) : q \in U\ q^1)\} \geqq f(q^2) > f(q^0)\}. \tag{5.28}$$

Then, it follows from (5.26) that the following relations must be met:

$$\begin{aligned} v(q^1) &= sup\{\ inf\ \{\ f(s) : s \in U\} : q^1 \in U \in \text{Ц}\}, \\ &\geqq inf\ \{f(q) : q \in U(q^1)\} \geqq f(q^2) \\ &> \mathrm{f}(q^0) \geqq v(q^0). \end{aligned} \tag{5.29}$$

In short, we have thus shown that q^1Fq^0 implies $v(q^1) > v(q^0)$, ensuring Property (5).

(Step 4) Property (1):

To prove Property (1), let us define the following new function:

$$\begin{aligned} w(q) \equiv \{x \in X : qx \leqq 1, \text{and } v(r) \geqq v(q) \\ \text{for any } r \in Q \text{ such that } rx \leqq 1\}. \end{aligned} \tag{5.30}$$

If we want to prove Property (1), then it suffices for us to show that for any $q \in Q$, $h\ (q) = w\ (q)$. First of all, from **Step 3** above, we clearly find $h\ (q) \subset w\ (q)$ for all $q \in Q$.

Now, by way of contradiction, let us suppose $x \in w\ (q) - h\ (q)$. Since $x \in X$, we have $x \in h\ (r)$ for some $r \in Q$, $r \neq q$. But we then obtain $q\ F\ r$, which implies $v\ (q) > v\ (r)$ by help of **Step 3** again. On the other hand, we have $r\,x = 1$, so that if $x \in w\ (q)$ then we must have $v\ (r) \geqq v\ (q)$, which gives a contradiction. Therefore, to get rid of a contradiction, we must have $h\ (q) = w\ (q)$ for all $q \in Q$.

(Step 5) Property (3):

To see Property (3), let $q^1, q^0 \in Q$ be such that $v(q^1) \geqq v(q^0)$ and $q^1 \neq q^0$, and let $q^t = (1-t)q^1 + t\,q^0$ for any $t \in (0, 1)$. Then, letting $x^t \in h\ (q^t)$, we find $(1-t)q^1x^1 + t\,q^0x^1 = q^tx^t = 1$, whence $q^1\,x^1 \leqq 1$ or $q^0x^1 \leqq 1$. On the one hand, if $q^1x^1 \leqq 1$, then we find $q^1\ F\ q^t$, implying that $v\ (q^1) > v\ (q^t)$ by means of **Step 3**. On the other hand, if $q^0x^1 \leqq 1$, then we find $q^0F\ q^t$, implying that $v\ (q^1) \geqq v\ (q^0) > v\ (q^t)$. In either case, we thus obtain, $v(q^1) > v\ (q^t)$. This ensures Property (3).

(Step 6) Property (4):

Finally, to prove Property (4), let $q^1, q^0 \in Q$ be such that $q^1 \leq q^0$. Then clearly, we obtain $q^1\ Fq^0$. In the light of **Step 3** above, this gives $v(q^1) > v(q^0)$. The proof is now complete.

Q.E.D.

Theorem 5.5 is a very important theorem, deriving the *indirect* utility function and its many properties. Property (1) indicates that for a given normalized-price vector q, the choice set $h(q)$ constitutes bundles x that *minimize* the *indirect* utility of normalized-price vectors r subject to the budget constraint $r\ x \leqq 1$. Property (2) means that for a given q, the "inferior" set $\{r: v(q) \geqq v\ (r)\}$ is closed in Q; namely, the *indirect* utility function is lower semi-continuous. Property (3) shows that v is strictly quasi-concave, whereas Property (4) says that v is monotonous. Finally, it follows from Property (5) that the revealed favorability relation F can be well represented by the *indirect* utility v. In other words, there is a sort of heritability relationship between revealed favorability and *indirect* utility.

Now, we are in a position to make a bridge between the *indirect* and *direct* utility functions. To carry out such a nice task, we find it necessary to introduce an additional *strong assumption* on the demand function h:

(D) For any $q \in Q$, h (q) is a singleton, or a set, which contains exactly one element.

Let us recall that the strong axiom ***(SF)*** of reveled favorability implies the unique invertibility of h. Therefore, if h satisfies both *(SF)* and ***(D)***, then h is a *one-to-one correspondence* between Q and X.

Let us define that a function g on X as follows:

$$\text{For any } x \in X, g(x) = h^1(x). \tag{5.31}$$

Then, it is clear that g $(h$ $(q)) = q$ for all $q \in Q$, and that h $(g$ $(x)) = x$ for all $x \in X$. Since g is the inverse of h on X, it immediately follows from Theorem 5.5 that for any $x \in X$, the following equation holds:

$$g(x) \\ = \{q \in Q : qx \leqq 1, \text{and } v(r) \geqq v(q) \text{for any } r \in Q \text{such that } rx \leqq 1\}. \tag{5.32}$$

Now, we are in a position to define a real-valued function (a *direct* utility function) u on X, i.e., that h uniquely *maximizes* u over B. More specifically, we would like to establish the following important theorem.

Theorem 5.6 (The Derivation of the *Direct* Utility) Suppose that the demand function h satisfies the budget assumption ***(H)*** and the strong axiom of revealed favorability ***(SF)*** *together with the strong demand assumption* **(D)**. Then, there exists a real-valued function u on Q, namely the *direct utility function*, such that the following series of properties hold.

1. (maximality) For any $q \in Q$,
   h $(q) = \{x$: $x \in b$ $(q) \cap X$, and u $(r) \geqq u$ (q) for any y
   such that $y \in b$ $(q) \cap X\}$.
2. (closeness) NOT DERIVABLE
3. (strict concavity) If u $(x^1) \geqq u$ (x^0), x^1, $x^0 \in X$, $x^1 \neq x^0$, and
   $x^t = (1 - t)x^1 + t\, x^0$, then u $(x^1) > u$ (x^t),
4. (monotonicity) If $x^1 \geq x^0$ and x^1, $x^0 \in X$, then u $(q^1) > u$ (q^0).
5. (heritability) For any x^1, $x^0 \in X$, x^1 H x^0 implies u $(q^1) > u^0$ (q^0).

Proof While the proof seems to be analogous to, yet not exactly as the same as, the proof of Theorem 5.5, it will carefully be carried out in a step-by-step fashion.

(Step 1) Property (4):

Under the strong axiom of revealed favorability ***(SF)*** and the strong demand assumption ***(D)***, the demand function h is a one-to-one correspondence between the normalized-price space Q and the range of the demand correspondence X. It is recalled that X need not be identical to the whole commodity space Y: X may be a proper subset of Y.

Letting $q, r \in Q, x = h(q)$, and $y = h(r)$, this immediately implies that $x\,H\,y$ if and only if $q\,F\,r$. Suppose that $x^1, x^0 \in X$ are such that $x^1\ H\ x^0$, and let $q^1 = g\ (x^1)$ and $q^0 = g\ (x^0)$. Then, we have $q^1\ F\ q^0$, so that $v\ (q^1) > v\ (q^0)$ by virtue of the last Theorem 5 (5) above. Consequently, we find the following:

$$u(x^1) = v(g(x^1)) > v(g(x^0)) = u(x^0). \tag{5.33}$$

This establishes Property (4).

(Step 2) Property (1):

To prove Property (1), for any $q \in Q$, let us define the following equation:

$$k(q) = \{x : x \in b(q) \cap X, \text{and } u(x) \geqq u(y) \text{for any } y \text{ such that } y \in b(q) \cap X\}. \tag{5.34}$$

Then evidently, it suffices to show that for any $q \in Q$, $h\ (q) = k\ (q)$. First of all, it is obvious by Property (4) that $h\ (q) \subset k\ (q)$. Next, by way of contradiction, assume $x \in k\ (q) - h\ (q)$. Then, $y = h\ (q)$ for some $y \neq x$. This yields $y\ H\ x$, so that $u\ (y) > u\ (x)$ by means of **Step 1** above. On the other hand, since $x \in k\ (q)$ and $y \in b(q) \cap X$, it follows Eq. (5.34) that $u\ (x) \geqq u\ (y)$, which is a contradiction. Therefore, to get rid of a contradiction, we must conclude that $h\ (q) = k\ (q)$ for all $q \in Q$.

(Step 3) Property (2):

To show Property (2), let us suppose that $u\ (x^1) \geqq u\ (x^0)$, $x^1, x^0 \in X$, $x^1 \neq x^0$, $t \in (0, 1)$, and $x^t = (1 - t)x^1 + t\,x^0 \in X$. Then we find $1 = g\ (x^t)x^t = (1 - t)\ g\ (x^t)x^1 + t\ g\ (x^t)x^0$. Hence, we have $g\ (x^t)\ x^1 \leqq 1$ or $g\ (x^0)\ x^0 \leqq 1$. On the one hand, if $g\ (x^t)x^1 \leqq 1$, then $x^t\ H\ x^1$, so that $u\ (x^1) > u\ (x^1) \geqq u\ (x^0)$ by Property (4) and hypothesis. On the other hand, if $g\ (x^t)\ x^0 \leqq 1$, then we have $x^1\ H\ x^1$ by Property (4) again. In either case, we thus obtain $u\ (x^1) > u\ (x^t)$, establishing Property (2).

(Step 4) Property (3):

Finally, to prove Property (3), let $x^1, x^0 \in X$ be such that $x^1 \geq x^0$. Then clearly, we have $x^1\ H\ x^0$, which gives $u\ (x^1) > u\ (x^0)$ by means of **Step 1** above.

Thus, the proof is now complete.

Q.E.D.

Comparison of the last Theorem 5.5 and the present Theorem 5.6 indicates the fundamental duality that exists between the *indirect* and *direct* utility functions.

1. First of all, minimizing the *indirect* utility v of normalized prices q is equivalent to the *direct* utility u of commodities x, with the identical budget constraint $qx \leqq 1$ being imposed in both instances.

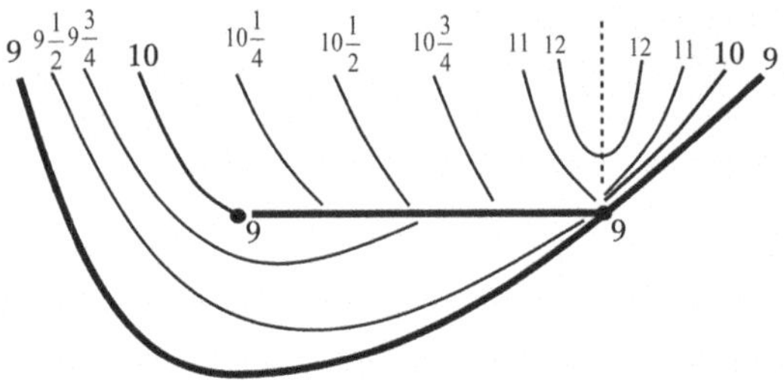

Fig. 5.18 Sonnenschein's ingenious example: (1) the set $\{x: u(x) \geqq 10\}$ is *not* closed; (2) the line segment 【9, 9】does *not* belong to the range of h either

2. Second, v is strictly quasi-convex on Q, whereas u is strictly quasi-concave on X. v is decreasing on Q, whereas u is increasing on X.
3. Third, v represents the *indirect* revealed favorability relation F on Q, whereas u represents the *direct* revealed preference relation H on.

It should be noticed, however, that such nice symmetry between the *indirect* and *direct* utilities is not perfect and may possibly break down under the present assumptions including ***(H)***, ***(SF)*** and ***(D)***.

Let us pay attention to Property (2) of the present Theorem 5.6, which is unfortunately not derivable under the present assumptions. Indeed, although the *indirect* utility function v is lower semi-continuous (see the last Theorem 5.5(2), the *direct* utility function u may not be upper semi-continuous under the present assumptions only.

Such an "inconvenient truth" is well illustrated by Sonnenschein's ingenious Example 3 (Chipman et al. 1971, pp. 274–275). As the proverb goes, seeing is believing here. Although Fig. 5.18 appears to be intentionally distorted by a charting device, we believe that it grasps the essence of things. There, the demand function h generated by the indicated *direct* utility function u clearly satisfies the assumptions ***(H)***, ***(SF)***, and ***(D)***, but no upper semi-continuous function can represent h: for instance, the set $\{x: u\ (x) \geqq 10\}$ is not closed.

It should be remarked that X, the range of h, is not convex in the Sonnenschein example: indeed, the line segment 【9, 9】 does not belong to the range of h. This once again clarifies the critical role played by the convexity assumption on the range of the demand function in exploring the exact theory of consumer's demand.

5.6 Choice and Rationality: Effectiveness and Limitations

In the above, we have been assumed that the consumer's choice behavior is *always rational and consistent*. In fact, the weak and strong axioms of revealed favorability on the normalized-price space and those of revealed preference on the commodity space may well represent such *rationality and consistency*. Besides, as was shown above, the *convexity of the range of the demand function* is also a good indicator of the *wise and sensible human judgment*. It is recalled that Richter (1971) discussed many possible kinds of the consumer's rationality including "transitive rational" and

" reflexive rational." In line with his way of argument, we may also add that the consumer is "convex rational" if his demand range is wide enough to satisfy the convex condition.

In the light of the long history of economic theory, there lies the academic struggle between "Econs" and "Humans." According to Richard H. Thaler (2015), our good friend and respected Rochester graduate, many standard models tend to use a fictional creature called *homo economics* or simply Econs. Econs are generally supposed in the majority of economic books and papers including Samuelson (1955, 7th edition 1967) and the present book per se. However, we have to be very careful of the possible danger of going too far or too much. If we may put it in the strongest terms, we are allowed to regard Econs as a sort of "rational fools" a la Amartia Sen (1987) in the sense that people rationally choose goods and services by following the weak and strong axioms of revealed favorability or those of revealed preference. Or equivalently, people are fictionally assumed to optimize their indirect or direct utilities.

In contrast, Humans are just human beings or *homo sapience.* Compared with fictional Econs, Humans are supposed to have a lot of non-rational feelings such as envy, hatred, optimism, pessimism, sympathy, compassion, and the like. In the world in which many Humans live, the traditional economic theory might be far from satisfactory. We need to establish a more comprehensive model of human behavior including a variety of complicated psychologies. As is stressed by George A. Akerlof and Robert J. Shiller (2009), a nice bridge between economic theory and human psychology must urgently be built.

It is true that getting economic theory back on its feet again will not an easy task, presumably requiring a very long arduous way ahead. We believe, however, that where there is a strong will, there is a nice way out.

References

Akerlof GA, Shiller RJ (2009) Animal spirits: how human psychology drives the economy, and why it matters for global capitalism. Princeton University Press, Princeton

Allen RGD (1936) Professor Slutsky's theory of consumer's choice. Rev Econ Stud 3:120

Chipman JS, Hurwicz L, Richter MK, Sonnenschein HF (eds) (1971) Preferences, utility, and demand. Harcourt, New York

Diewert WE (1982) Duality approaches to microeconomic theory. In: Arrow KJ, Intriligator MD (eds) Handbook of mathematical economics. Elsevier, New York

Hicks J (1939 2nd ed 1946), Value and capital: an inquiry into some fundamental principles of economic theory. Oxford University Press

Hotelling H (1932) Edgeworth's taxation paradox and the nature of demand and supply functions. J Polit Econ 40:571–616

Houthakker H (1950) Revealed preference and the utility function. Economica 17:159–174

Hurwicz L, Richter MK (1971) Revealed preference without demand continuity assumptions. In: Chipman JS, Hurwicz L, Richter MR, Sonnenschein HF (eds) Preferences, utility, and demand: a Minnesota symposium. Harcourt Brace Jovanovich, New York

Kuga K (1969) Revealed preference theory and the indirect utility function. Discussion paper 49, Institute of Social and Economic Research, Osaka University

Lau LJ (1969) Duality and the structure of utility functions. J Econ Theory 1:374–396

McKenzie LW (1957) Demand theory without a utility index. Rev Econ Stud 24:185–189

Richter MK (1966) Revealed preference theory. Econometrica 34:635–645

Richter MK (1971) Rational choice. In: Chipman JS, Hurwicz L, Richter MK, Sonnenschein HF (eds) Preferences, utility, and demand: a Minnesota symposium. Harcourt Brace Jovanovich, New York

Sakai Y (1977) Revealed favorability, indirect utility, and direct utility. J Econ Theory 14(1): 113–129

Samuelson PA (1947) Foundations of economic analysis. Harvard University Press

Samuelson PA (1955, 7th edition 1967) Economics: an introductory analysis. McGraw-Hill, New York

Sen AK (1987) On ethics and economics. Blackwell, Oxford

Slutsky E (1915) Sulla teoria del bilancio del consumatore (in Italian) [English translation: on the theory of consumer's behavior]. Giornale degli Economisti

Thaler RH (2015) Misbehaving: the making of behavioral economics. Norton, New York

Uzawa H (1960, revised 1971) Preferences and rational choice in the theory of consumption. Contained in Arrow KJ, Karlin S, Suppes P (eds) Mathematical methods in the social sciences 1959, Stanford: University Press; Revised paper (1971) is in: Chipman JS, Hurwicz L, Richter MR, Sonnenschein HF (eds) Preferences, utility, and demand: a Minnesota smposium. Harcourt Brace Jovanovich, New York

Weddepohl HN (1970) Axiomatic choice models and duality. Rotterdam University Press, Rotterdam

Yokoyama T (1953) A logical foundation of the theory of consumer's demand. Osaka Econnomic Papers 2:71–79

Chapter 6
Producer Decision and Input Demand Theory: An Axiomatic Approach

Abstract This chapter is concerned with an axiomatic approach to input demand theory. By help of general production possibility sets, we intend to derive decomposition equations in input demand theory, which have been rather neglected so far in the economics literature. Special attention is paid to important comparison between the firm's expansion effect and the consumer's income effect. We discuss the question of how and to what extent the expansion effect is distinct from the income effect. In this connection, the LeChatelier–Samuelson principle is also discussed.

Keywords Producer decision · Comparative static analysis · Expansion effect · Inferior input · LeChatelier–Samuelson principle

6.1 The Firm's Expansion Effect Versus the Consumer's Income Effect

The purpose of this chapter is to rigorously analyze producer decision and production possibility sets, thus exploring the *axiomatic foundations* of production theory. In older days, Hicks (1946, 1953), Samuelson (1947), Morishima (1953a, b), and others elaborated upon the well-established analysis of profit-maximizing firm's demand for inputs. Their main mathematical tools were *classical differential calculus.*

In more recent times, however, there have emerged a series of papers that bravely study the *axiomatic foundations of production theory* on the basis of *more modern*

This chapter is a completely revised version of Sakai (1973). I am grateful to Lionel McKenzie (Rochester), James Friedman (Rochester), William Brock (Chicago), Michio Morishima (London), Akira Takayama (Perdue), Akihiro Amano (Kobe), Jiichi Hayashi (Kobe), Edward Diewart (British Columbia), Trout Rader (Missouri), Asatoshi Maeshiro (Pittsburgh), Yoshihiko Ohtani (Perdue), and Masayoshi Hirota (Tokyo Keizai) for valuable suggestions and constant encouragement.

Y. Sakai, *Games, Decisions, and Markets*, New Frontiers in Regional Science: Asian Perspectives 74, https://doi.org/10.1007/978-981-99-5285-4_6

topological tools. For this point, see Scott (1962), Bear (1965), Ferguson (1966, 1968, 1969), Rader (1968), Basett and Borcheding (1969), Hirota and Sakai (1969), Syrquin (1970), Shephard (1970), Arrow and Hahn (1971), Sakai (1973, 1974, 1975), and Malinvaud (1985). After the 1990s until the present day, unifying the axiomatic foundations of both consumption and production theories, brand new approaches named *duality approaches to microeconomic theory* have emerged and widely flourished, with Diewert (1982, 2018), Varian (1999, 2009), and McKenzie (2002) being eminent accomplishments.

A number of those economists abovementioned have bravely attempted to clear up a matter of long-standing confusion concerning the analogy between *consumer's income effect* and *producer's* expansion *effect.* While consumer demand theory and input demand theory appear to be analogous, they are not the same at all. Even a small difference at start may produce a big distinction in goal. The question of how and to what extent they are really different is our main concern here.[1]

As was independently shown by Hirota and Sakai (1969) and also by Syrquin (1970), the effect of a change in the price of a certain input on the demand for another input can be divided into the following two separate effects. They are *a substitution effect along the old isoquant* and *an expansion effect along the new expansion path.* Unfortunately, the mathematical tool they employed was simple calculus, thus lacking mathematical rigor and fineness. The main purpose of this chapter is to make our input demand theory mathematically more sophisticated than the previous attempts, yet developing it in many possible applications to economic reality.

As will be seen below, we obtain a sort of decomposition equation in input demand theory, which appears to correspond well to the famous Slutsky equation in consumer demand theory. The correspondence between the two decomposition equations, however, is not quite exact: indeed, they look similar but are not identical. We must pay special attention to the critical difference between the word "similar" and the word "identical."[2]

In this chapter, we want to widely apply the powerful method of McKenzie (1957), which was originally used for the development of consumption theory, to the new area of input demand theory. Then, we can successfully derive various kinds of decomposition equations. And in so doing. we develop the *new idea of a compensated change in output price when a certain input price varies.* This idea is seemingly analogous to, but not exactly the same as, the familiar concept of a

[1]In the 1960s and the 1970s, I myself was a graduate student at the University of Rochester. And completing my Ph.D. thesis, I became an assistant professor at the University of Pittsburgh. In hindsight, I was so fortunate to be surrounded by many outstanding professors such as Lionel McKenzie, Edward Zabel, Ronald Jones, Hugh Rose, James Friedman, and Akira Takayama, and by many brilliant students including Jerry Green, Jose Sheinkman, Masayoshi Hirota, and Michihiro Ohyama. Though perhaps too late, I would like to say my sincere thanks to all of them.

[2]Historically speaking, the fundamental difference between consumer demand theory and input demand theory was first noticed by Hicks (1946), Chap. 7, and mathematically sophisticated by Samuelson (1947), Chap. 4. It seems, however, that the *similarity* between them was not fully developed by these authors, resulting in later development of input demand theory.

compensated change in income when a certain commodity price varies in the traditional consumption theory a la Hicks (1946).

So far, input demand theory has been developed in connection with the problem of *inferior inputs*. An input is called *normal* (or *inferior*) if a rise in output price causes an increase (or a decrease) in the demand in that input. We can obtain the following results. (1) While it is possible that all inputs are normal, it is not possible that they are all inferior. (2) In case a certain input is inferior, it is not possible that all other inputs are gross complements with it although it is possible that they are all gross substitute for it.

The contents of this chapter are as follows. Section 6.2 is addressed to a system or a production technology. In Sect. 6.3, the definition of cost and profit functions will be given, with a careful discussion of their properties. The topological approach of Shephard (1970) to the duality principle between cost and production is developed in these two sections. The total effect of a change in the price of a certain input on the demand for another input is decomposed into substitution and scale effects. In Sect. 6.4, we are first concerned with the properties of substitution and total effects and examine the question of how and to what extent they are similar to, or different from, those of substitution and total effects in consumption theory. The problems of inferior inputs and of net and gross substitutability are also discussed. Section 6.5 is devoted to various types of decomposition equations in input demand theory—one finite increment and two differential versions. It is rigorously shown that the demand curve of an input is not positively sloping, and the substitution and expansion effects always go in the same direction. In Sect. 6.6, we will apply our analysis to show the validity of the LeChatelier–Samuelson principle in input demand theory.

6.2 Production Possibility Sets

We are concerned with a firm that is faced with the problem of producing a single output from a combination of a finite number of inputs subject to a production technology. Let us suppose that there are n inputs and that input–output prices are competitively determined in the market, being independent of the firm's individual behavior.

In what follows, we will make full use of a powerful topological method. For this method, see McFadden (1966), Malinvaud (1985), Mas-Colell et al. (1995). McKenzie (2002), and Mitra and Nishimura (2009).

Let us denote an input bundle by $x = (x_1, x_2, ..., x_n)$. The set of all conceivable input bundles is denoted by X. X is the set of all nonnegative n-vectors:

$$X = \{x = (x_1, x_2, \ldots, x_n) : x \geqq 0\}. \tag{6.1}$$

Let us assume that for any $x \in X$, the largest output is conceivable and conveniently denoted by a *production function* $f(x)$. Let Y be the range of the production function:

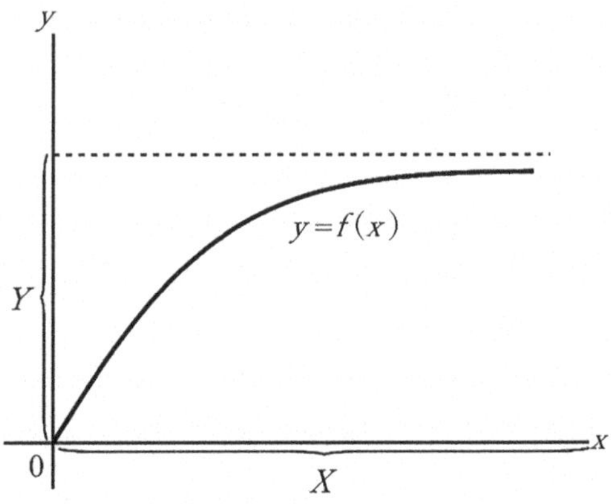

Fig. 6.1 A simple case of a single input and a single output

$$Y = \{y : y \geqq 0 \text{ and } y = f(x) \text{ for some } x \in X\}. \tag{6.2}$$

We assume that Y is nonempty, convex, and open above. Y need not be the nonnegative real line. For a simple case with a single input and a single output, X and Y are illustrated in Fig. 6.1.

We now define the following sets:

$$W = \{w = (w_1, w_2, \ldots, w_n) : w > 0\}, \tag{6.3}$$

$$P = \{p : p > 0\}. \tag{6.4}$$

Evidently, W is the input price space or the set of all input vectors, and P is the output price space or the set of all conceivable output prices.

There is an alternative useful description of the production function $f(x)$. To show this, let us define the following *production possibility sets*:

$$\bar{A}(y) = \{x : x \in X \text{ and } f(x) \geqq y\}, \tag{6.5}$$

$$\underline{A}(y) = \{x : x \in X \text{ and } f(x) \leqq y\}, \tag{6.6}$$

$$I(y) = \bar{A}(y) \cap \underline{A}(y) = \{x : x \in X \text{ and } f(x) = y\}. \tag{6.7}$$

It is noted that the *upper possibility set* $\bar{A}(y)$ is the set of all input vectors that are capable of producing *at least* the output y, whereas the *lower possibility set* $\underline{A}(y)$ is the set of all input vectors that are capable of producing *at most* the output y. The set $I(y)$, which is the intersection of $\bar{A}(y)$ and $\underline{A}(y)$, is clearly the familiar *isoquant frontier* corresponding to y.

For a simple case with two inputs, x_1 and x_2, those three production possibility sets, $\bar{A}(y)$, $\underline{A}(y)$ and $I(y)$, are well illustrated in Fig. 6.2.

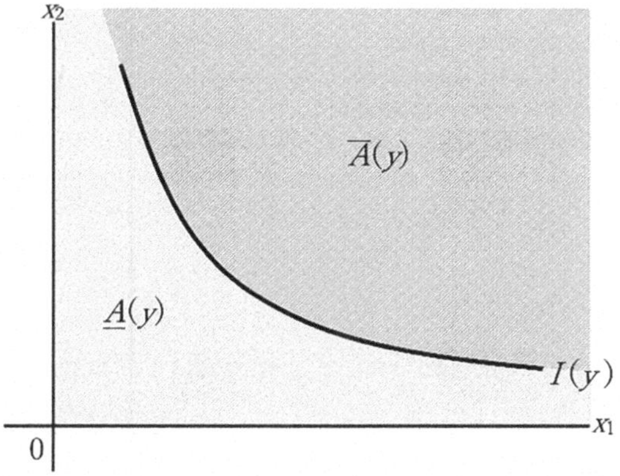

Fig. 6.2 Three production possibility sets: $\bar{A}\ (y)$, $\underline{A}(y)$, and $I\ (y)$

We are ready to define a *production technology* as a family of those production possibility sets satisfying the following four assumptions[3]:

Assumption (A1) If $0 \in I\ (y)$, then $y = 0$.

Assumption (A2) For each $y \in Y$, $\bar{A}\ (y)$ and $\underline{A}\ (y)$ are nonempty and closed in X.

Assumption (A3) Let us take $x^0 \in \bar{A}\ (y^0)$ and $x^1 \in \bar{A}\ (y^1)$ where $x^0 \neq x^1$. For any $t \in (0, 1)$, if we put $x^t = (1 - t)x^0 + tx^1$ and $y^t = (1 - t)y^0 + t\,y^1$, then we must find $x^t \in \mathrm{Int}\,\bar{A}\ (y^t)$ or the interior of the set $\bar{A}\ (y^t)$.

Assumption (A4) If $x^1 \geqq x^0$ and $x^0 \in \bar{A}\ (y)$, then $x^1 \in \bar{A}\ (y)$.

These assumptions require detailed explanations. (A1) states that a positive output cannot be obtained from a null input bundle. In short, nothing comes from nothing. (A2) asserts that any output level y is attainable for some input bundle x, and that the production function $f(x)$ is continuous. Therefore, there should exist neither gaps nor jumps on the production curve.

Remarkably, (A3) together with (A2) clearly implies that the set $\bar{A}\ (y)$ is a "strictly convex set," meaning that for any x^0 and x^1, namely two points of $\bar{A}\ (y)$, if $x^t = (1 - t)x^0 + tx^1$, then x^t must belong to the interior of the set $\bar{A}\ (y)$, not to the boundary $I\ (y)$. The essence of (A3) is well illustrated in Fig. 6.3. In terms of the production function $f(x)$, the production curve is a "strictly concave curve," with "no flat boundary."

And finally, (A4) ensures that free disposal of inputs is possible. In other words, the producer can dispose of any extra inputs with no costs. Although this may not necessarily reflect the reality, it is theoretically a very convenient assumption.

[3]For a topological and convex approach to production theory, see Uzawa (1964), Nikaido (1968), Shephard (1970), and others.

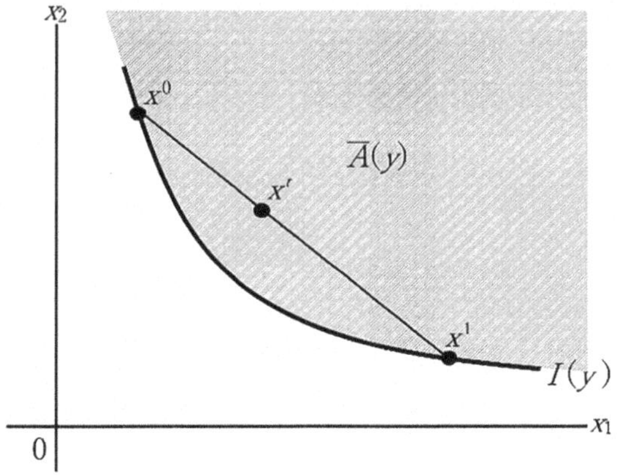

Fig. 6.3 The strict convexity of the upper production set $\bar{A}$ (y). It is noted that x^t belongs to the interior of $\bar{A}$ (y)

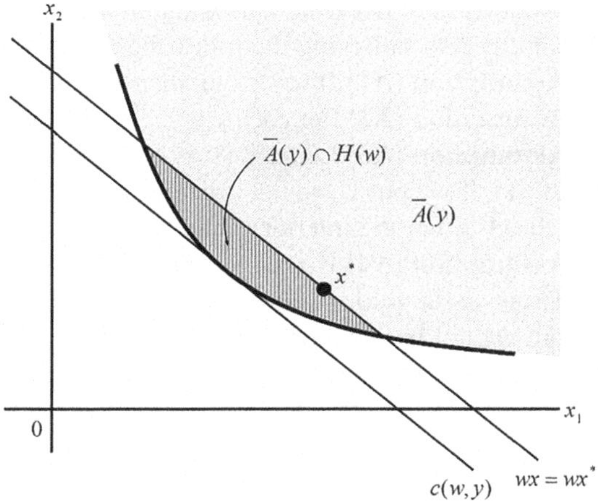

Fig. 6.4 The cost function is illustrated here

6.3 The Cost and Profit Functions: Definitions and Properties

In what follows, we assume that Assumptions (A1)–(A4) are met for the production technology. In this section, the definition of cost and profit functions will be given and their properties will carefully be investigated.

First of all, for any $(w, y) \in W \times Y$, we define a *cost function* as follows:

$$c\ (w, y) = \mathbf{Min}\ \{w\, x : x \in \bar{A}\ (y)\}. \tag{6.8}$$

To see that the cost function $c\ (w, y)$ is well defined, we take any arbitrary $x^* \in \bar{A}$ (y) and let $H\ (w) = \{x{:}\ x \in X$ and $wx \leqq wx^*$. Since $\bar{A}\ (y)$ is closed by (A2) and $H(w)$ is clearly compact, it follows that $\bar{A}\ (y) \cap H\ (w)$ is compact as well. Therefore, as is seen in Fig. 6.4, the continuous function wx takes on a minimum on $\bar{A}\ (y) \cap H\ (w)$.

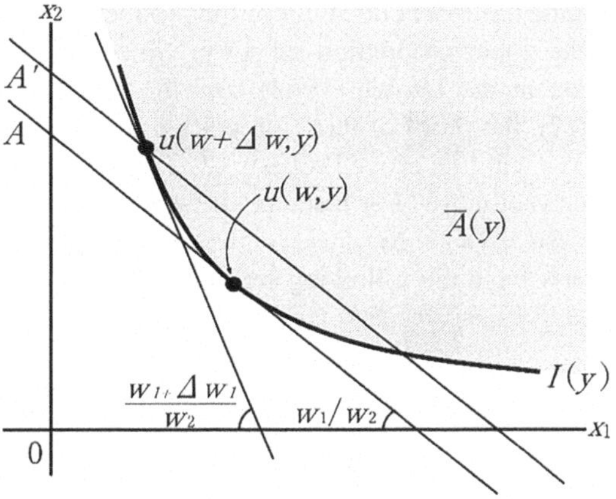

Fig. 6.5 For the simple case of two inputs, we clearly have the following: $c(w_1, w_2, y) = w_1 \cdot u_1(w_1, w_2, y) + w_1 \cdot u_2(w_1, w_2, y) \leqq w_1 \cdot u_1(w_1 + \Delta w_1, w_2, y) + w_1 u_2(w_1 + \Delta w_1, w_2, y)$. $A'B'$ lies above AB, so that we find $w \cdot u(w, y) < w \cdot u(w + \Delta w, y)$

Note that for any $x \in \bar{A}(y) - H(w)$, we have $wx > wx^*$. Hence, the function wx attains a minimum on $\bar{A}(y)$.

In connection with the cost function $c(w, y)$, we define a *compensated input function* as follows:

$$u(w,y) = \{x : x \in \overline{A}(y) \text{ and } wx = c(w,y)\}. \tag{6.9}$$

The newly defined function $u(w, y)$ indicates the input bundle demanded by the firm when input prices are w and output is to be y. Under Assumptions (A1)–(A4), $u(w, y)$ is the unique element of X minimizing wx subject to $x \in \bar{A}(y)$. Consequently, we should have the following equation:

$$c(w,y) = w \cdot u(w,y). \tag{6.10}$$

The relation between $c(w, y)$ and $u(w, y)$ is very important and clearly seen in Fig. 6.5. We are now in a position to establish the following useful lemma:

Lemma 6.1 (Properties of $c(w, y)$ and $u(w, y)$) For all $(w, y) \in W \times Y$, we have the following properties:

1. $c(w, y)$ is continuous in (w, y).
2. $c(w, y)$ is differentiable in w, and $\partial c(w, y)/\partial w_i = u_i(w, y)$ for $i = 1, 2, ..., n$.
3. $c(w, y)$ is homogeneous of degree one in w.
4. $c(w, y)$ is concave in w.
5. $c(w, y)$ is strictly convex in y.
6. If $y^0 > y^1$, $y^0, y^1 \in Y$, then $c(w, y^0) > c(w, y^1)$.
7. $u(w, y)$ is continuous in (w, y).
8. $u(w, y)$ is homogeneous of degree zero in w.

Proof We note that the present assumptions (A1)–(A4) are slightly stronger than the assumptions used by Uzawa (1964) and also by Friedman (1972): They did not assume the *strict* concavity of the production function $f(x)$, but merely the concavity of the upper production set $\bar{A}$ (y). Since the proof of Properties (3), (4), and (6) is found in the Uzawa (1964), and the proof of Property (1) is found in Friedman (1972), the proof of those properties are omitted here.

We will first prove Property (2). Let us pay attention to a small change of the input price vector Δw. For instance, if we have $\Delta w = (\Delta w_1, 0, \ldots, 0)$, then we must have $w + \Delta w = (w_1 + \Delta w_1, w_2, \ldots, w_n)$. In the light of Eq. (6.10) above, it is noted that we clearly have the following set of equations:

$$\begin{aligned} &c(w+\Delta w, y) - c(w, y) \\ &\quad = (w+\Delta w)u(w+\Delta w, y) - w \cdot u(w, y). \\ &\quad = w【u(w+\Delta w, y) - u(w, y)】 + \Delta w \cdot u(w+\Delta w, y). \end{aligned} \tag{6.11}$$

$$c\ (w, y) = w \cdot u\ (w, y) < w \cdot u\ (w+\Delta w, y). \tag{6.12}$$

$$\begin{aligned} c(w+\Delta w, y) &= (w+\Delta w)u(w+\Delta w, y) \\ &\leqq (w+\Delta w)u(w, y) = c(w, y) + \Delta w \cdot u(w, y). \end{aligned} \tag{6.13}$$

Note here that for the simple case of two inputs, the validity of Eq. (6.12) is graphically illustrated in Fig. 6.5.

Taking advantage of Eqs. (6.11) and (6.12), we can derive the following:

$$\begin{aligned} &【c(\mathrm{w}+\Delta w, y) - c(w, y)】/|\Delta w| \\ &= w【u(w+\Delta w, y) - u(w, y)】/|\Delta w| + \Delta \mathrm{w} \cdot \mathrm{u}(\mathrm{w}+\Delta \mathrm{w}, \mathrm{y})/|\Delta \mathrm{w}| \\ &\geqq \Delta w \cdot u(w+\Delta w, y)/|\Delta w|. \end{aligned} \tag{6.14}$$

In the light of Eq. (6.13), we find the following:

$$\begin{aligned} &【c(w+\Delta w, y) - c(w, y)】/|\Delta w| \\ &\leqq \Delta w \cdot u(w, y)/|\Delta w|. \end{aligned} \tag{6.15}$$

Now, if we take care of Eqs. (6.14) and (6.13), then we obtain the following:

$$\begin{aligned} 0 &\leqq 【\mathrm{c}(w+\Delta w, y) - c(w, y) - \Delta w \cdot u(w+\Delta w, y)】/|\Delta w| \\ &\leqq \Delta w【u(w, y) - u(w+\Delta w, y)】/|\Delta w|. \end{aligned} \tag{6.16}$$

Since u (w, y) is continuous by Property (1), we can see that u (w, y) $-$ u $(w$ $+$ $\Delta w,$ $y)$ $\rightarrow$ 0 as Δw $\rightarrow$ 0. Because Δw $/|\Delta w|$ is clearly bounded, the right-hand side of Eq. (6.16) goes to zero as Δw $\rightarrow$ 0. Therefore, we must have the following:

$$\begin{aligned}&\mathbf{Lim}_{\Delta w\to 0}[c(w+\Delta w,y)-c(w,y)-\Delta w\cdot u(w,y)]/|\Delta w|\\&\quad=\mathbf{Lim}_{\Delta w\to 0}[c(w+\Delta w,y)-c(w,y)-\Delta w\cdot u(w+\Delta w,y)]/|\Delta w|=0.\end{aligned}\tag{6.17}$$

This clearly shows differentiability of c (w, y) with respect to w. Now, let us specify $\Delta w = (0, ..., 0, \Delta w_i, 0, ..., 0)$. Then, it is easily seen that the following equation holds:

$$\partial c\ (w,y)/\partial w_i = u_i\ (w,y) \text{ for } i=1,2,\ldots,n. \tag{6.18}$$

To prove Property (5), let us take $y^0, y^1 \in Y$, $y^0 \neq y^1$, and let $y^t = (1-t)y + t\,y^1$, $t \in (0, 1)$. In the light of Assumptions (A2) and (A3), we find $(1-t)u\ (w, y^0) + tu(w, y^1) \in \operatorname{Int} \bar{A}\ (y^t)$. Hence, we must have the following:

$$\begin{aligned}&(1-t)c(w,y^0)+tc(w,y^1)\\&\quad=w[(1-t)u(w,y^0)+tu(w,y^1)]>c(w,y^t).\end{aligned}\tag{6.19}$$

To show Property (7), let us take a sequence $\{(w^k, y^k)\}$ in $W \times Y$ such that

$$\mathbf{Lim}_{k\to\infty}\left(w^k,y^k\right)=(w,y)\in W\times Y. \tag{6.20}$$

Then, by definition, we have $c\ (w^k, y^k) = w^k{\cdot}u\ (w^k, y^k)$. Since $c\ (w^k, y^k)$ is continuous by Property (1), letting $k \to \infty$ yields the following:

$$c\ (w,y)=\mathbf{Lim}_{k\to\infty}\, c\left(w^k,y^k\right)=w\cdot\mathbf{Lim}_{k\to\infty}\, u\left(w^k,y^k\right). \tag{6.21}$$

Since $u\ (w, y)$ is uniquely determined for (w, y), we must obtain the following:

$$\mathbf{Lim}_{k\to\infty}\, u\ (w^k,y^k)=u\ (w,y). \tag{6.22}$$

This clearly shows that $u\ (w, y)$ is continuous in (w, y).

Finally, to see Property (8), let $\lambda > 0$. Since $c\ (w, y)$ is homogeneous of degree one in w by Property (3), we must have the following:

$$\lambda w\cdot u\ (\lambda w,y)=c\ (\lambda w,y)=\lambda\cdot c\ (w,y)=\lambda w\cdot u\ (w,y). \tag{6.23}$$

Thus, we must obtain $u\ (\lambda w, y) = u\ (w, y)$, showing that $u\ (w, y)$ is homogeneous of degree zero in w.

Q.E.D.

At first glance, the contents of Lemma 6.1 appear to be rather technical and even too mathematical. It should be noticed, however, that it discusses many important properties of the cost and compensated input functions, which play critical roles in

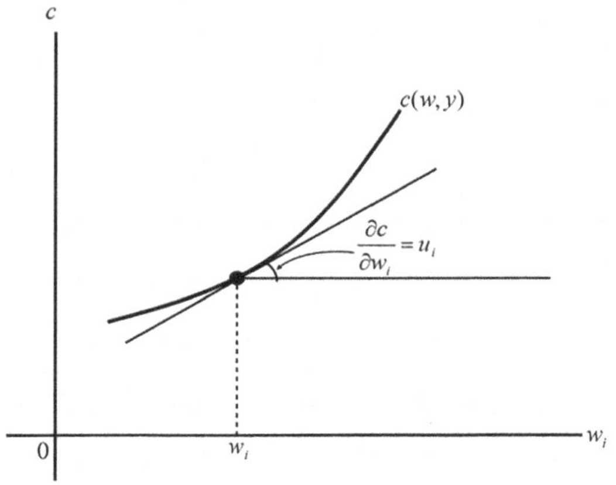

Fig. 6.6 A nice relation between c and u: $\partial c/\partial w_i = u_i$

the theory of cost and production. According to Properties (1), the cost function c (w, y) is continuous in (w, y), so that the *cost curve* as a graphical expression of the cost function is overall smooth and has neither gaps nor jumps throughout. Property (2) tells us that focusing on w only, c (w, y) is extremely smooth, having no kinks at all. Moreover, it also indicates a very nice bridge between the two functions, namely the cost function c (w, y) and the compensated input function u (w, y). Exactly speaking, For any i, we should have $\partial c\,(w, y)\,/\partial w_i = u_i\,(w, y)$, demonstrating that the rate of change of total cost when the price of an input changes is equal to the amount of compensated demand for that input. This delicate relation between the two functions is well illustrated in Fig. 6.6.

According to Properties (3), (4), (5), and (6), while the cost function c (w, y) is homogeneous one and concave in w, it is strictly convex in y and also strictly increasing. Such concave and convex relations should be worthy of attention. Finally, in the light of Properties (7) and (8), whereas the compensated input demand function u (w, y) is continuous in (w, y), it is homogeneous of degree zero in w only. This teaches us that u (w, y) reflects the nice properties of u (w, y).

In addition to Assumptions (A1)—(A4), we are now ready to postulate the following assumption.

Assumption (A5) Y is bounded from above.[4]

As will be seen later, the purpose of Assumption (A5) is to force the profit function to be defined later for any $(w, p) \in W \times P$. The newly added Assumption (A5) together with the previous assumptions on Y implies that $Y = [0, y^*)$ for some $y^* < +\infty$. Let $E\,(w, p) = \{x: x \in X \text{ and } pf\,(x) - wx \geqq 0\}$. Then, under Assumptions (A1)–(A5), the newly defined set E (w, p) is clearly bounded and closed; therefore, the continuous function $pf\,(x) - wx$ takes on maximum on $E\,(w, p)$.

[4]The purpose of Assumption (A5) is to force the profit function $\pi(w, p) = \{p{\cdot}f\,(x) - wx: x \in X\}$ to be defined for any $(w, p) \in W \times P$.

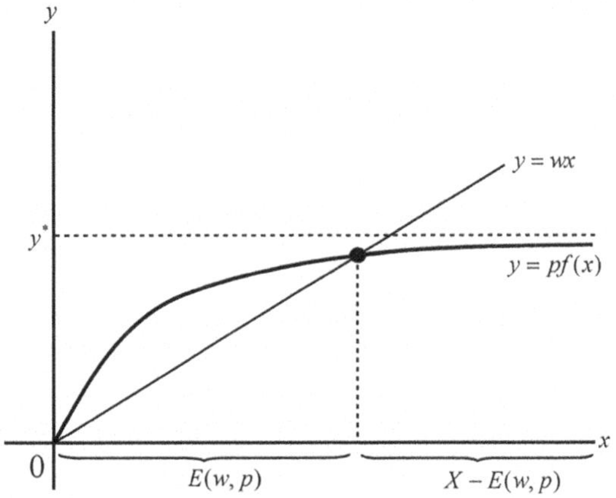

Fig. 6.7 The relationship between E (w, p) and X

Note that by Assumption (A1), $pf(x) - wx = 0$ for $x = 0$; $pf(x) - w\,x < 0$ for any $x \in X - E\ (w, p)$. Thus, the maximum attained on $E\ (w, p)$ is actually the maximum on the whole X.

Figure 6.7 clearly illustrates the relation between the set $E\ (w, p)$ and X.

Now, for any $(w, p) \in W \times P$, let us define a *profit function* $\pi(w, p)$ as follows:

$$\pi(w,p) = \text{Max}\ \{pf\ (x) - w\,x : x \in X\}. \tag{6.24}$$

Clearly, for any given price vector (w, p), $\pi(w, p)$ stands for the maximum profit, which can be obtained from the production technology. Since $f(x)$ is strictly concave by Assumption (A3), it is clear that $\pi(w, p)$ is positive for some $(w, p) \in W \times P$. Thus, an *optimum input function* and $x\ (w,\ p)$ and an *optimum output function* y $(w, p) \in W \times P$ are defined for all (w, p). Evidently, $x\ (w, p)$ and $y(w, p)$ respectively indicate the input bundle as follows.

$$x\ (w,p) = \{x : x \in X \text{ and } \pi(w,p) = pf\ (x) - w\,x\}. \tag{6.25}$$

$$y(w,p) = f\ (x\ (w,p)). \tag{6.26}$$

Evidently, $x\ (w, p)$ and $y\ (w, p)$ respectively indicate the input bundle demanded by the firm and the output bundle supplied by the firm. It is noted that the input–output vector $(x(w,\ p),\ y(w,\ p))$ is uniquely determined for each price vector $(w, p) \in W \times P$ since the profit $\pi(w, p)$is strictly concave in x.

The properties of the profit function along with those of the optimum input–output function will be seen in the following lemma.

Lemma 6.2 (Properties of $\pi(w, p)$) For all $(w, p) \in W \times P$, we have following properties:

1. $\pi(w, p)$ is differentiable in (w, p), and

$$\partial\pi(w,p)/\partial w_i = -x_i\ (w,y) \text{ for } i = 1, 2, \ldots, n,$$
$$\partial\pi(w,p)/\partial p = y\ (w,y).$$

2. $\pi(w, p)$ is homogeneous of degree one in (w, p).
3. $\pi(w, p)$ is convex in (w, p).
4. If $w^0 \geqq w^1 > 0$, then $\pi(w^0, p) \leqq \pi(w^1, p)$, with strict inequality if $x\ (w^0, p) > 0$.
5. If $p^0 > p^1 > 0$, then $\pi(w, p^0) > \pi(w, p^1)$, with strict inequality if $y\ (w^0, p^1) > 0$.
6. $x\ (w, p)$ and $y\ (w, p)$ are continuous in (w, p).
7. $x\ (w, p)$ and $y\ (w, p)$ are homogeneous of degree zero in (w, p).

Proof

(a) For convenience, let us first show that $\pi(w, p)$, $x\ (w, p)$, and $y\ (w, p)$ are all continuous in (w, p).

To this end, take a sequence $\{(w^k, p^k)\}$ in $W \times P$ such that the following equation holds:

$$\mathbf{Lim}_{k\to\infty}\ (w^k, p^k) = (w, p) \in W \times P.$$

In the light of Eq. (6.24), we should have the following:

$$\begin{aligned}\pi(w,p) &= p\ y(w,p) - w\ x(w,p)\\ &\geqq p\ y(w^k,p^k) - w\ x(w^k,p^k),\end{aligned} \tag{6.27}$$

$$\begin{aligned}\pi(w^k,p^k) &= p^k y(w^k,p^k) - w^k x(w^k,p^k)\\ &\geqq p^k y(w,p) - w^k x(w,p^k).\end{aligned} \tag{6.28}$$

Because of Assumption (A5), it is clear that the sequences $\{\pi(w, p)\}$, $\{x\ (w, p)\}$, and $\{y\ (w, p)\}$ are all bounded and contain convergent subsequences. Without loss of generality, we may assume that the original sequences themselves are convergent. By Eqs. (6.27) and (6.28), we derive the following:

$$\begin{aligned}\pi(w,p) &= p\ y(w,p) - w\ x(w,p)\\ &\geqq p\mathbf{Lim}_{k\to\infty} y(w^k,p^k) - w\mathbf{Lim}_{k\to\infty} x(w^k,p^k)\\ &= \mathbf{Lim}_{k\to\infty} p^k \mathbf{Lim}_{k\to\infty} y(w^k,p^k) - \mathbf{Lim}_{k\to\infty} w^k \mathbf{Lim}_{k\to\infty} x(w^k,p^k)\\ &= \mathbf{Lim}_{k\to\infty} \pi(w^k,p^k)\\ &\geqq \mathbf{Lim}_{k\to\infty} p^k y(w,p) - \mathbf{Lim}_{k\to\infty} w^k x(w,p^k)\\ &= p\ y(w,p) - w\ x(w,p).\end{aligned} \tag{6.29}$$

Therefore, we find $\mathbf{Lim}_{k \to \infty}\pi(w^k, p^k) = \pi(w, p)$, showing continuity of $\pi(w, p)$. Further, in the light of Eq. (6.29), we also obtain the following:

$$p\mathbf{Lim}_{k\to\infty}y(w^k, p^k) - w\mathbf{Lim}_{k\to\infty}x(w^k, p^k) \\ = p\, y(w,p) - w\, x(w,p). \tag{6.30}$$

By the uniqueness property of $x\ (w, p)$ and $y\ (w)$, it follows from Eq. (6.29) that $\mathbf{Lim}_{k \to \infty}\ y\ (w^k, p^k) = (w, p)$ and $\mathbf{Lim}_{\ k \to \infty}\ x\ (w^k, p^k) = x\ (w, p)$, thus assuring Property (6).

(b) By help of (a), we will next prove Property (1). In the light of Eqs. (6.28) and (6.29), we find, for any change $(\Delta w, \Delta p)$ in the price vector, we obtain the following equations.

$$\pi(w,p) \geqq p \cdot y(w + \Delta w, p + \Delta p) - w \cdot x(w + \Delta w, p + \Delta p) \\ = \pi(w + \Delta w, p + \Delta p) + \Delta w \cdot x(w + \Delta w, p + \Delta p) \\ - \Delta p \cdot y(w + \Delta w, p + \Delta p), \tag{6.31}$$

$$\pi(w,p) \geqq (p + \Delta p)y(w,p) - (w + \Delta w)x(w,p) \\ = \pi(w,p) - \Delta w \cdot x(w,p) + \Delta p \cdot y(w,p). \tag{6.32}$$

In the light of Eqs. (6.31) and (6.32), we can easily find the following:

$$- \Delta w \cdot x(w,p) + \Delta p \cdot y(w,p) \leqq \pi(w + \Delta w, p + \Delta p) - \pi(w,p) \\ \leqq - \Delta w \cdot x(w + \Delta w, p + \Delta p) + \Delta p \cdot y(w + \Delta w, p + \Delta p).$$

Therefore, we can derive the following inequalities.[5]

$$0 \leqq \frac{\pi(w + \Delta w, p + \Delta p) - \pi(w,p) + \Delta w \cdot x(w,p) - \Delta p \cdot y(w,p)}{|(\Delta w, \Delta p)|} \\ \leqq \frac{- \Delta w[x(w + \Delta w, p + \Delta p) - x(w,p)] + \Delta p\, y(w + \Delta w, p + \Delta p) - y(w,p)]}{|(\Delta w, \Delta p)|} \tag{6.33}$$

In the light of (a) above, $[x\ (w + \Delta w,\ p + \Delta p) - x\ (w,\ p)] \to 0$ and $[y\ (w + \Delta w,\ p + \Delta p) - y\ (w,\ p)] \to 0$ as $(\Delta w,\ \Delta p) \to 0$. Since $\Delta w\ /|(\Delta w,\ \Delta p)|$ and $\Delta p\ /|(\Delta w,\ \Delta p)|$ are bounded, the extreme right-hand side of Eq. (6.33) goes to zero as $(\Delta w,\ \Delta p) \to 0$. Therefore, we can derive the following.

---

[5] According to the definition of the length of a vector, we find the following: $|(\Delta w,\ \Delta p)| = [\sum_i (\Delta w_i)^2 + (\Delta p)\ ^2]^{1/2}$.

$$\mathrm{Lim}_{(\Delta w, \Delta p) \to 0}\left(\frac{\pi(w + \Delta w, p + \Delta p) - \pi(w, p) + \Delta w \cdot x(w, p) - \Delta p \cdot y(w, p)}{|(\Delta w, \Delta p)|}\right)$$
$$= 0, \tag{6.34}$$

assuring differentiability of $\pi(w, p)$.

Letting $(\Delta w, \Delta p) = (0, ..., 0, \Delta w_i, 0, ..., 0)$, Eq. (6.34) yields the following.

$$\partial\pi(w,p)/\partial w_i = -x_i\ (w,p) \text{ for } i = 1, 2, \ldots, n.$$

Similarly, letting $(\Delta w, \Delta p) = (0, ..., 0, \Delta p)$, it yields the following.

$$\partial\pi(w,p)/\partial p = y\ (w,p).$$

(c) To see Properties (2) and (7) together, let $\pi > 0$. Then, we find the following.

$$\begin{aligned}\lambda p \cdot y(\lambda w, \lambda p) - \lambda w \cdot x(\lambda w, \lambda p) &= \pi(\lambda w, \lambda p) \\ &= \mathbf{Max}\{\lambda p\, f(x) - \lambda w\, x : x \in X\} \\ &= \lambda\mathbf{Max}\{p\, f(x) - w\, x : x \in X\} \\ &= \lambda\pi(w,p) = \lambda p \cdot y(w,p) - \lambda w \cdot x(w,p).\end{aligned}$$

By the uniqueness property of $x\ (w, p)$ and $y\ (w, p)$, we find $x\ (\lambda w, \lambda p) = x\ (w, p)$ and $y\ (\lambda w, \lambda p) = y\ (w, p)$, thus assuring (7). We also have $\pi(\lambda w, \lambda p) = \lambda\pi\ (w, p)$, thereby assuring (2).

(d) To see (3), select $(w^0, p^0), (w^1, p^1) \in W \times P$, $(w^0, p^0) \neq (w^1, p^1)$. Let $(w^t, p^t) = (1 - t)(w^0, p^0) + t\ (w^0, p^0)$, $t \in (0, 1)$. Then, we have the following.

$$\begin{aligned}\pi(w^t, p^t) &= p^t y(w^t, p^t) - w^t x(w^t, p^t) \\ &= (1 - t)\{p^0 y(w^t, p^t) - w^0 x(w^t, p^t)\} + t\{p^1 y(w^t, p^t) - w^1 x(w^t, p^t)\} \\ &\leqq (1 - t)\pi(w^0, p^0) + t\pi(w^1, p^1)\end{aligned}$$

Clearly, this assures Property (3).

(e) To see Property (4), let $w^0, w^1 \in W$, $w^0 \geq w^1$, and $p \in P$. Then, we have the following.

$$\begin{aligned}\pi(w^0, p^t) &= p \cdot y(w^0, p) - w^0 \cdot x(w^0, p) \\ &\leqq p \cdot y(w^0, p) - w^1 \cdot x(w^0, p),\end{aligned}$$

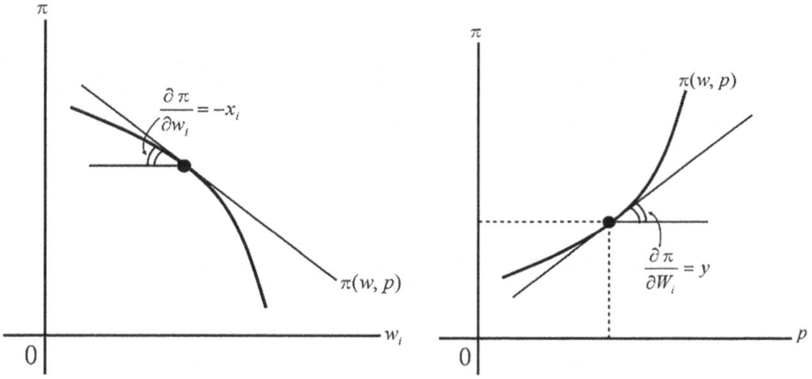

Fig. 6.8 Nice relationship between profit and input-output: $\partial\pi/\partial w_i = -x_i$ and $\partial\pi/\partial p = y$

with strict inequality if $x\,(w^0, p) > 0$. This assures Property (4). The proof of Property (3) should be parallel to the above.

Q.E.D.

Lemma 6.2 is quite interesting in making a bridge between the profit function and the optimum input–output function. The essence of Property (1) is expressed by the two equations; $\partial\pi/\partial w_i = -x_i\,(w, p)$ and $\partial\pi/\partial p = y\,(w, y)$. Graphically, it can easily be seen in Fig. 6.8.

In plain English, on the one hand, the rate of change of total profit when the price of an input price changes is equal to the amount of the input demanded by the firm, multiplied by (−1). On the other hand, the rate of change of total profit when the price of an output changes is equal to the amount of the output supplied by the firm.

According to Property (1), the profit function $\pi(w, p)$ is not only continuous but also differentiable in (w, p), so that the *profit curve* as a graphical expression of the profit function is overall very smooth, with no kinks, having neither gaps nor jumps throughout. Property (2) implies that when (w, p) doubles, π also doubles. It follows from Property (3) that for any $(w, p) \in W \times P$, and for any fraction $t \in (0, 1)$, the following inequality holds:

$$\begin{aligned}&(1-t)\pi(w^0,p^0) + t\,\pi(w^1,p^1)\\&>\pi((1-t)w^0 + t\,w^1, (1-t)p^0 + t\,p^1).\end{aligned} \tag{6.35}$$

According to Properties (3) and (4), while π tends to decrease as w rises, but it tends to increase as p rises, thus agreeing with common sense.

Finally, by help of Properties (6), inputs and outputs continuously change in response to a small change in (w, p). And Property (7) shows us that when all input prices and all output prices change in the same proportion, all inputs and all outputs are expected to remain unaffected, agreeing our common sense.

Now, we are ready to establish the following interesting lemma.

Lemma 6.3 (Marginal Cost and Output Price) For all $(w, p) \in W \times P$, we obtain the following:
$\partial c\ (w, y\ (w, p))/\partial y = p$, almost everywhere.

Proof By the definition of $c\ (w,\ y\ (w,\ p))$, we note that the following inequality must hold:

$$p \cdot y(\mathrm{w},p) - c(w,y(\mathrm{w},p)) \\ \geqq py - c(w,y) \text{for all} (w,y) \in W \times Y. \tag{6.36}$$

Since $c\ (w,\ y)$ is concave in y by Lemma 6.1 (4), it must be almost everywhere differentiable in y.[6]

Further, Eq. (6.36) asserts that the function $p\ y - c\ (w,\ y)$ is maximized at $y = y\ (w,\ p)$. Therefore, the first-order characterization for the maximum yields the following.

$$\partial (p\ y - c\ (w,y))/\partial y = 0, \text{almost everywhere.}$$

This implies that $p - \partial c\ (w,\ y)/\partial y = 0$ at $y = y\ (w, p)$ almost everywhere, from which immediately $\partial c\ (w,\ y\ (w,\ p))/\partial y = p$ almost everywhere.
Q.E.D.

Lemma 6.3 is an important lemma, saying that the marginal cost of an output is equal to output price. Surely, it demonstrates profit-maximizing behavior of the firm from a different angle. The essence of Lemma 6.3 is graphically illustrated in Fig. 6.9.

We are now in a position to discuss how the optimum and compensated input functions are related with each other.

Lemma 6.4 (Properties of x (w, p) and y (w, p)) For all $(w, p) \in W \times P$, we obtain the following equations:

1. $x\ (w,\ p) = u\ (w,\ y\ (w,\ p))$.
2. $c\ (w,\ y\ (w,\ p)) = w{\cdot}x\ (w,\ p)$.
3. $\pi(w,\ p) = p{\cdot}y\ (w,\ p) - c\ (w,\ y\ (w,\ p))$.

Proof To see Property (1), we first note $x(w, p) \in \bar{A}\ (y\ (w, p))$. Let us next take $x^s \in \bar{A}\ (y\ (w, p))$, and $x^s \neq x\ (w, p)$. This is possible by Assumptions (A2), and (A4). Now, let a "middle point" $x^t = (1 - t)x(w, p) + tx^s$, $t \in (0, 1)$. Then, by means of Assumptions (A2) and (A3), we obtain $x^t \in \text{Int}\ \bar{A}\ (y\ (w,\ p))$, so that $wx^t > w{\cdot}x\ (w,\ p)$ as is well illustrated in Fig. 6.10.

[6]For the properties of convex and concave functions, see Fenchel (1953), Rockafellar (1970), Shephard (1970), and Varian (1999, 2009).

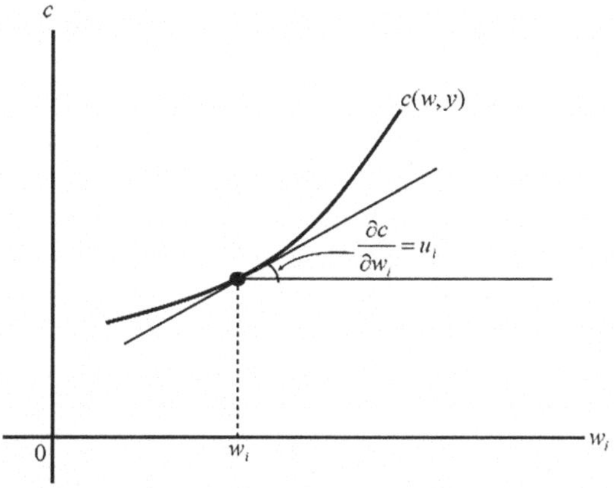

Fig. 6.9 At equilibrium, the following relation holds almost everywhere: $\partial c(w, y)/\partial y = p$ at $y = y(w, p)$

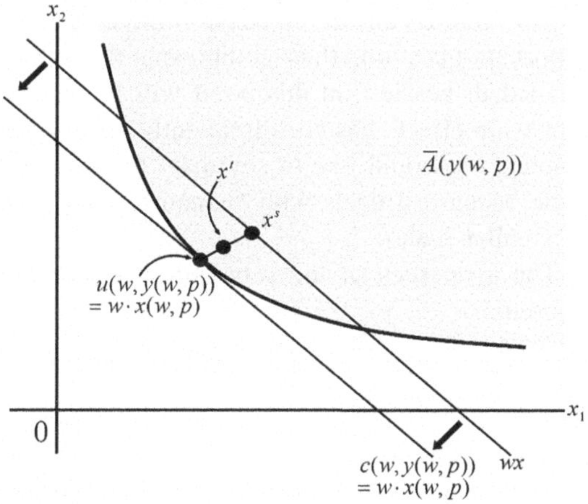

Fig. 6.10 The proof of Lemma 6.10 (1) is graphically illustrated here

Letting $t \to 0$ yields $x^t \to x(w, p)$, and $wx \geqq w \cdot x(w, p)$. Since $x(w, p) \in \bar{A}(y(w, p))$, and $u(w, y(w, p))$ is uniquely determined for each (w, p), it must follows that $u(w, y(w, p)) = x(w, p)$. This proves Property (1).

From (1), we find $c(w, y(w, p)) = w \cdot x(w, p) = u(w, y(w, p))$, and $\pi(w, p) = p \cdot y(w, p) - c(w, y(w, p))$, assuring Properties (2) and (3).

Q.E.D.

From this lemma, the economic significance of $u(w, y(w, p))$ is quite clear. For an arbitrary price vector (w, p), the optimum input–output combination $(x(w, p))$ is uniquely determined from the production technology. Then, $c(w, y(w, p))$ denotes the cost level that governs the output level $y(w, p)$, and $u(w, y(w, p))$ is the corresponding input bundle. It is also seen that the profit gained by the firm is equal to the difference between $p \cdot y(w, p)$ and $c(w, y(w, p))$.

6.4 Properties of the Substitution and Total Effects

In the previous section, we sought the information about conditions governing inputs demanded by the firm at given input–output prices. In this section, we are now ready to use it to discover how the inputs will change when these prices vary.

Now, let us consider a change in w_j. Then, we will see that its impact on x_i can be divided into the following two effects: a *substitution effect* along the old isoquant and an *expansion effect* (or a *scale effect*) along the new expansion path (or scale path). Although the naming of the first effect is fairly traditional, that of the second effect might be quite new. The composition of the total effect into the substitution and the expansion (or scale) effect in production theory was first noticed by Hirota and Sakai (1969), and later developed by Sakai (1973, 1974), and others. Such composition in production seems to be similar to, yet not as the same as the composition into the substitution and income effects in consumption theory. How and to what extent the expansion effect differs from the income effect is a very important question, thus composing the main theme of the present chapter. A detailed discussion on this point will be made in the next section. Although "the expansion effect" has been frequently used in the previous literature and also in this chapter, we would like to say that "the scale effect" may be an equally appealing name because it deals with "expansion" (or larger scale) as well as "contraction" (or smaller scale).

The properties of the substitution effect can be summarized in the following theorem.

Theorem 6.5 (Differential Properties of u (w, y)) For all $(w, y) \in W \times Y$, we have the following properties:

1. $(\partial u_i/\partial w_1)w_1 + \ldots + (\partial u_i/\partial w_n)w_n = 0$ for any i, almost everywhere.
2. $w_1 (\partial u_1/\partial w_j) + \ldots + w_n (\partial u_n/\partial w_j) = 0$ for any j, almost everywhere.
3. $\partial u_i/\partial w_j = \partial u_j/\partial w_i$ for any i, j, almost everywhere.
4. The following matrix is negative semi-definite, almost everywhere.

$$\begin{pmatrix} \partial u_1/\partial w_1 & \cdots & \partial u_1/\partial w_n \\ \vdots & & \vdots \\ \partial u_n/\partial w_1 & \cdots & \partial u_n/\partial w_n \end{pmatrix}.$$

Proof Recalling that c (w, y) is concave in w by Lemma 6.1 (4), it should be *almost everywhere twice differentiable* in w. Note that by Lemma 6.1 (2), we find that ∂c $(w, y)/\partial w_i = u_i$ (w, y) for all i. Therefore, it follows that for any i, u_i (w, y) is almost everywhere differentiable in w.[7]

[7]Note that any concave or convex function is *almost everywhere twice differentiable*. See Alexandroff (1939).

Fig. 6.11 $wu\ (w^s, y) \geqq wu\ (w, y)$ for all $w^s \in W$

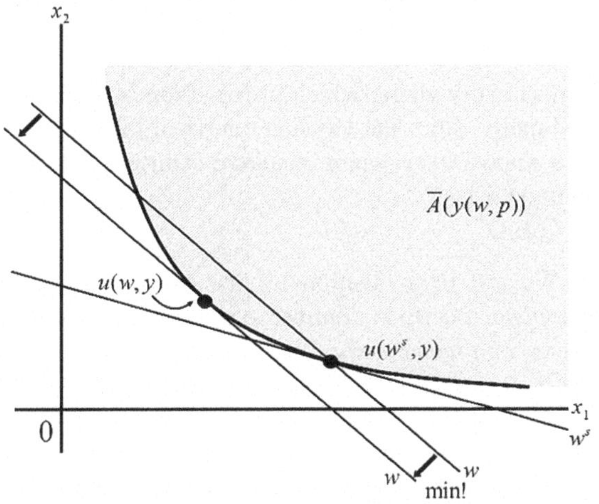

With those preparations in mind, we also recall that $u\ (w, y)$ is homogeneous of degree zero in w by Lemma 6.1 (8). Here, if we apply the famous Euler theorem on homogeneous equations, we immediately find the following equation.

$$(\partial u_i/\partial w_1)\ w_1 + \ldots + (\partial u_i/\partial w_n)\ w_n = 0 \text{ for any } i, \text{almost everywhere,}$$

assuring Property (1).

To prove Property (2), it is noted that by definition of the compensated function u, the following inequality holds:

$$w \cdot u\ (w^s, y) \geqq w \cdot u\ (w, y) \text{ for all } w^s \in W. \tag{6.37}$$

Note that the validity of Eq. (6.37) is well illustrated in Fig. 6.11. Eq. (6.37) tells us that the function $w{\cdot}u\ (w, y)$ attains minimum among all the functions of the form $w{\cdot}u\ (w^s, y)$ at $w^s = w$.

Hence, the first-order characterization for the minimum yields the following.

$$\partial(w \cdot u\ (w^s, y))/\partial w^s{}_j = 0 \text{ at } w^s = w, j = 1, \ldots, n,$$

almost everywhere, from which Property (2) follows.

To see Property (3), we recall the well-known Young theorem on differential calculus, which says that whenever a function is twice differentiable, the order of differentiation is not important.[8]

Therefore, by Lemma 6.1 (2), for any i, j, we obtain the following.

[8] See Takagi (1961), pp. 57–58.

$$\partial u_i/\partial w_j = \partial^2 c/\partial w_j \partial w_i = \partial^2 c/\partial w_i \partial w_j = \partial u_j/\partial w,$$

almost everywhere, which proves Property (3).

Finally, since the Hessian matrix of the concave function $c\,(w, y)$ with respect to w is almost everywhere negative semi-definite, Property (4) is easily derived from Lemma 6.1 (2).

Q.E.D.

We are in a position to discuss a new concept of *net substitution and net complementarity* in contrast to a more traditional concept of *gross substitution and gross complementarity.* We say that the input i is a *net substitute for the input j* if $\partial u_i/\partial w_j > 0$, and a *net complement for it if* $\partial u_i/\partial w_j < 0$. Possibly as a rare case, we could have the special situation in which $\partial u_i/\partial w_j \leqq 0$ *for all i.* Generally speaking, when the price of a certain input price rises, the demand for that input tends to decrease if output is to be constant. In other words, the compensated demand curve tends to be negatively sloping. Coupled with this result, Property (1) of Theorem 6.5 implies that whereas it is possible that all other inputs are net substitutes for an input, it is not possible that they are net complements for it. In short, although net substitutability can be seen everywhere, net complementarity is a rare phenomenon. Property (2) also shows further limits on the possibility of net complementarity. Property (3) asserts that the substitution effect should be symmetrical between two inputs. Naturally, this agrees with common sense.

The properties of the total effect will be summarized in the following theorem.

Theorem 6.6 (Differential Properties of $x\,(w, p)$ and $y\,(w,p)$) For all $(w, p) \in W \times P$, we have the following properties:

1. $(\partial x_i/\partial w_1)w_1 + \ldots + (\partial x_i/\partial w_n)\, w_n + (\partial x_i/\partial p)\, p = 0$ for any i, almost everywhere.
2. $(\partial y/\partial w_1)w_1 + \ldots + (\partial y/\partial w_n)\, w_n + (\partial y/\partial p)\, p = 0$, almost everywhere.
3. $\partial x_i/\partial w_j = \partial x_j/\partial w_i$ for all i, j, almost everywhere.
4. $\partial x_i/\partial p + \partial y/\partial w_i = 0$ for all i, almost everywhere.
5. The following matrix is positive semi-definite, almost everywhere.

$$\begin{pmatrix} -\partial x_1/\partial w_1 & \cdots & -\partial x_1/\partial w_n & \partial x_1/\partial p \\ \vdots & & \vdots & \vdots \\ -\partial x_n/\partial w_1 & \cdots & -\partial x_n/\partial w_n & \partial x_n/\partial p \\ -\partial y/\partial w_1 & \cdots & -\partial y/\partial w_n & \partial y/\partial p \end{pmatrix}.$$

Proof We first note that the profit function $\pi(w, p)$ is almost everywhere twice differentiable since it is convex by Lemma 6.2 (3). Since $\partial\pi/\partial w_i = -x_i$ for any i, and $\partial\pi/\partial p = y$, this obviously implies that both input function $x(w, p)$ and the output function $y(w, p)$ are almost everywhere differentiable. Such differential properties are quite important for further derivations.

From Lemma 6.2 (7), we see that $x(w, p)$ and $y(w, p)$ are homogeneous of degree 1. Therefore, by applying the famous Euler theorem on homogeneous functions here, we can immediately obtain Properties (1) and (2).

Next, let us recall the famous Young theorem, which says that if a function is twice differentiable, the order of differentiation does not matter, leading to the same result. So, if we apply the Young theorem to the profit function $\pi(w, p)$, which is convex by Lemma 6.2 (7), then we can obtain the following equations almost everywhere:

$$\partial x_i/\partial w_j = -\partial^2\pi/\partial w_j\partial w_i = -\partial^2\pi/\partial w_i\partial w_j = \partial x_j/\partial w_i \text{ for all } i, j,$$

$$\begin{aligned}\partial x_i/\partial p + \partial y/\partial w_i &= -\partial^2\pi/\partial p\,\partial w_i + \partial^2\pi/\partial w_i\partial p\\ &= -\partial^2\pi/\partial w_i\,\partial p + \partial^2\pi/\partial w_i\partial p = 0 \text{ for all } i.\end{aligned}$$

Therefore, Properties (3) and (4) are surely assured.

Finally, since the risk function $\pi(w, p)$ is convex, its Hessian matrix must be positive semi-definite, almost everywhere. We note that the following set of equations hold.

$$\partial^2\pi/\partial w_j\,\partial w_i = \partial x_i/\partial w_j,\ \partial^2\pi/\partial p\partial w_i = \partial x_i/\partial p,$$

$$\partial^2\pi/\partial w_j\,\partial p = \partial y/\partial w_j,\ \partial^2\pi/\partial p^2 = \partial y/\partial p.$$

Hence, Property (5) is definitely assured.

Q.E.D.

To see the economic significance of Theorem 6.6, let us introduce several useful concepts here. We say that the input i is a *gross substitute for the input j* if $\partial x_i/\partial w_j > 0$; a *gross complement for it if* $\partial x_i/\partial w_j < 0$. We also say that the input i is a *normal input* if $\partial x_i/\partial p > 0$; an *inferior input* if $\partial x_i/\partial p < 0$. If we employ these concepts, we can give new light on the relationship between inputs and outputs on the one hand and input prices and output prices on the other hand.

First of all, Property (4) of Theorem 6.4 tells us that the "supposedly normal situation" under which $\partial x_i/\partial w_i \leqq 0$ for any i and $\partial y/\partial p \geqq 0$ is certainly plausible but not inevitable. As can easily be expected, while the input demand curve tends to be negatively sloping, the output supply curve tends to be positively sloping. Property (4) means that $\partial x_i/\partial p > 0$ (or <0) if and only if $\partial y/\partial w_i < 0$ (or >0). In plain English, thus means that the input i is a normal input (or am inferior input) if and only if a fall in the price of the input leads to an increase (or a decrease) in output.

In the light of Property (5), we find $\partial y/\partial p \geqq 0$. So, Property (2) implies that although it is likely that all of inputs are normal, it is a mission impossible that they are all inferior. Now suppose that a certain input, say x_i, is an inferior input, so that $\partial x_i/\partial p < 0$. It follows from Property (1) that, in such a case, it is not possible at all that all other inputs are gross complements for the input i although it is truly possible that they are all gross substitutes for it. Because of Property (3), the *total* effect is symmetrical between two inputs. However, we already know from Theorem 6.5 (3) above that the *substitution* effect is symmetrical between them. Therefore, the *expansion* effect, as the difference of the total and substitution effects, must be symmetrical as well.[9]

Finally, in the light of Properties (3) and (5), we obtain the following inequalities for all $i \neq j$.

$$(\partial x_i/\partial w_i)\ (\partial x_j/\partial w_j) \geqq (\partial x_i/\partial w_j)^2.$$

$$(-\partial x_i/\partial w_i)\ (\partial y/\partial p) \geqq (\partial x_i/\partial p)\ (-\partial y/\partial w_1).$$

Clearly, these inequalities indicate dominance of the *own effects* over the *cross effects.* Summing up, the differential properties of $x\ (w,\ p)$ and $y\ (w,\ p)$ are so important that they will be utilized for further discussions on input demand theory.

6.5 Decomposition Equations in Input Demand Theory

As stated above, the *total effect* of a change in the price of an input on the demand for another input can be split up into the two separate effects, namely the *substitution and scale effects.* The purpose of this section is to use the previous results to derive various types of *decomposition equations in input demand theory.* Our attention will be mainly devoted to seeing how and to what extent they are analogous to, or distinct from, the famous Slutsky equations in consumer demand theory (see Slutsky 1915; McKenzie 1957).

Let us attempt to decompose the change of the demand for the input *i* responding to a change in the price of the input *j*. For that purpose, let us consider the following increment.

$$\Delta x_i\ (w,p) = x_i\ \left(w + \Delta_j\, w, p\right) - x_i\ (w,p), \text{with } \Delta_j w = \left(0,\ \ldots, 0, \Delta w_j, 0,\ \ldots, 0\right).$$

Then, by Lemma 6.4 (1) above, we obtain the following.

$$\begin{aligned}\Delta x_i(w,p) &= u_i\left(w + \Delta_j w, y\left(w + \Delta_j w, p\right)\right) - u_i(w, y(w,p)).\\ &= u_i\left(w + \Delta_j w, y(w,p)\right) - u_i(w, y(w,p))\\ &\quad + u_i\left(w + \Delta_j w, y\left(w + \Delta_j w, p\right)\right) - u_i\left(w + \Delta_j w, y(w,p)\right).\end{aligned} \tag{6.38}$$

Now, let us newly define the quantities SE_{ij} and EE_{ij} as follows.

[9]Note that the expansion effect, which is always symmetrical, markedly contrasts with the traditional income effect, which may be *non -symmetrical.*

$$SE_{ij} = [u_i\left(w + \Delta_j w, y\left(w, p\right)\right) - u_i\left(w, y\left(w, p\right)\right)]/\Delta w_j.$$

$$EE_{ij} = [u_i\left(w + \Delta_j w, y\left(w + \Delta_j w, p\right)\right) - u_i\left(w + \Delta_j w, y\left(w, p\right)\right)]/\Delta w_j.$$

Then, it is easily seen from Eq. (6.38) that the following equation holds.

$$\Delta x_i / \Delta w_j = SE_{ij} + EE_{ij}. \tag{6.39}$$

In our opinion, Eq. (6.39) has very important implications. It is noted that SE_{ij} and EE_{ij} respectively show the *substitution effect* and the *expansion effect.* At first appearance, it seems to be the *finite increment version* of the famous Slutsky equation in consumer demand theory. In fact, a change Δw_j in the price of the input j affects the behavior of the firm in two different ways. While it causes a change in input price ratios, which induces technical substitution among inputs *along the old isoquant,* it entails a change in the profit maximizing output *along the new scale path.*

On the one hand, the first substitution effect stands for the *variation in the optimum combination of inputs within the isoquant class to which the original x (w, p) belongs.* Then, $c\ (w + \Delta_j w, y\ (w, p))$ represents the corresponding level of cost, which *keeps the firm within the same isoquant class as before the input price change, even in the new input price situation $w + \Delta_j w$.* On the other hand, the scale effect represents the *shift of the optimum input combination in the new input price situation $w + \Delta_j w$, responding to the change of output level from the old level $w + \Delta_j w$ to the new level $y(w + \Delta_j w, p)$.*

These two distinct effects are well illustrated in Fig. 6.12. There, the movement from Point A to Point B indicates the substitution effect, and the movement from Point B to Point C demonstrates the expansion effect.

It is recalled that Eq. (6.39) merely stands for the *difference version of the decomposition equation.* We are now ready to derive its *differential version,* which is perhaps more interesting than the difference version. To this end, it is necessary to make an additional assumption as follows.

Assumption (A6) For all $(w, p) \in W \times P$, $x(w, p)$ and $y(w, p)$ are differentiable in (w, p).

This newly added assumption (A6) is not so far from the previous assumptions (A1)—(A5). In fact, as was seen above, the latter ones already assure *almost everywhere differentiability of $x(w, p)$* and $y(w, p)$. The only difference between the previous assumptions and the newly added assumption comes down to the difference between "*almost* everywhere" and "everywhere." Moreover, when (A6) is assumed, it is not hard to see from Lemma 6.4 that $u(w, y)$ and $c(w, y)$ are also differentiable.

Now, we are in a position to establish and prove one of the most important theorems in this chapter.

Theorem 6.7 (The First Kind of Decomposition Equations) For all $(w, p) \in W \times P$, we have the following eqs.

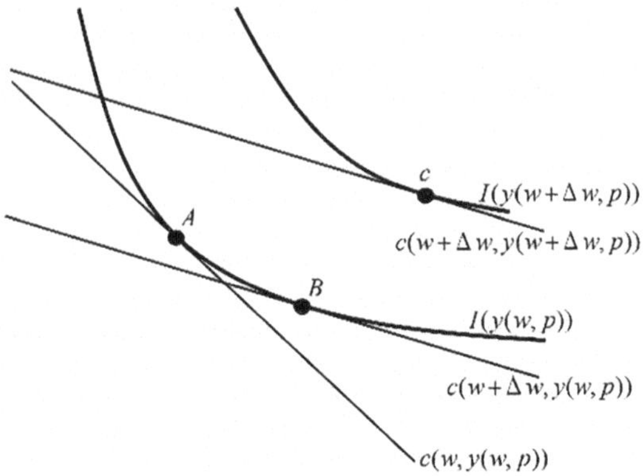

Fig. 6.12 The substitution and expansion effects are graphically illustrated

1. $\partial x_i/\partial w_j = \partial u_i/\partial w_j + (\partial u_i/\partial y)\ (\partial y/\partial w_j)$ for all i, j.
2. $\partial x_i/\partial p = (\partial u_i/\partial y)\ (\partial y/\partial p)$ for all i.

Proof In the light of Lemma 6.4 (1), the proof of (1) and (2) is easy and straightforward. Indeed, we know that the following important equation holds.

$$x\ (w, p) = u\ (w, y\ (w, p)).$$

If we apply the well-known rule on the differentiation of a composite function, then we can immediately obtain the desired Properties (1) and (2).

Q.E.D.

Property (1) indicates the *first version* of decomposition equation in input demand theory in *differential terms.* Exactly speaking, we find that $\partial x_i/\partial w_j = \partial u_i/\partial w_j + (\partial u_i/\partial y)\ (\partial y/\partial w_j)$. The first term $(\partial u_i/\partial w_j)$ on the right-hand side tells us to what extent a change in w_j influences u_i. The second term $(\partial u_i/\partial y)\ (\partial y/\partial w_j)$ on the right-hand, representing the remarkable scale effect term, shows the *composite effect* containing the following two partial derivative terms:

1. The partial derivative term $(\partial y/\partial w_j)$, representing the change in the optimum output corresponding to the change in the price of the input j,
2. The partial derivative term $(\partial u_i/\partial y)$, showing the change in the optimum input responding to the variation in the output abovementioned.[10]

Property (2) demonstrates that the total effect of a change in input price on the input i can be decomposed into the following two partial derivative terms:

[10]The *first* version of decomposition equation was merely referred to by Basett & Borcherding (Basett and Borcheding 1969) without any proof and later was given a formal proof by Hirota and Sakai (1969) and Syrquin (1970) by using simple calculus. It is noted that the most exact form of decomposition equations in input demand theory is given here in terms of the compensated input functions. Also see Okuguchi et al. (1989).

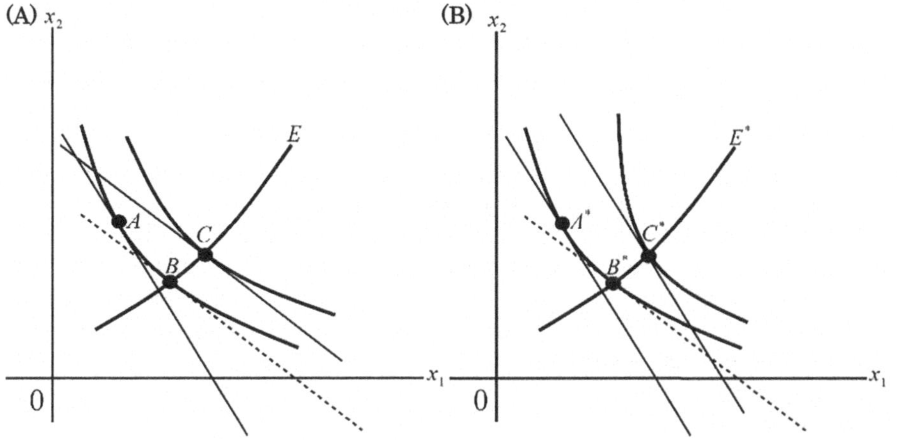

Fig. 6.13 Decomposition in consumer demand vs. decomposition in input demand: Namely, (A → B → C) (**A**) vs. (A* → B* → C*) (**B**)

1. The partial derivative term ($\partial y/\partial p$), representing the change in the optimum output as a result of the change in output price,
2. The partial derivative term ($\partial u_i/\partial y$), showing that the change in the optimum input corresponding to the variation in output above mentioned.

At first appearance, the combination of the firm's substitution and expansion effects seems to be analogous that of the consumer's well-known substitution and income effects. We must say, however, that *they are not exactly the same.*

Whereas in consumption theory we have simply to consider the relations between those commodities, which can be regarded as being essentially similar, in production theory we have *two different kinds of commodities to consider—inputs and outputs.* Therefore, their mutual relations and their cross-relations take a little more disentangling. We remind readers that, to obtain a decomposition equation in consumption theory, it is necessary to imagine a compensated change in consumer's income so as to keep the same preference level as before, in spite of a change in the price of a certain commodity. In contrast to such a compensated change in income in consumption theory, in production theory we have to introduce the idea of *a compensated change in output price,* which was first introduced by Hirota and Sakai (1969) and later developed by Sakai (1973), Otani (1982), Diewert (1982, 2018), and others. More exactly speaking, when the price of a certain input price varies, we imagine such a compensated change in output price as would induce the firm to maintain the output level as before the input price change.

The similarity and difference between decomposition in consumer demand and that in input demand are well illustrated in Fig. 6.13. In Panel (A), the movement of point from A to B and that from B to C respectively indicate the substitution effect and the income effect in consumer demand decomposition. In Panel (B), similar movements $A^* \rightarrow B^*$ and $B^* \rightarrow C^*$ respectively show the substitution effect and the expansion effect in input demand decomposition. On the one hand, in Panel (A), when p_1 falls, consumer's budget line twists from a_1a_2 to c_1 c_2. Since consumer's income stays fixed in spite of a fall in p_1, the two point a_2 and c_2 on the vertical axis must coincide. On the other hand. in Panel (B), when w_1 falls, producer's cost line shifts from $a_1{*}a_2{*}$ to c_1 ${*}c_2{*}$. Note that there are no fixed budget lines in input demand. Since producer's cost has to change responding to a fall in w_1, the two points $a_2{*}$ and $c_2{*}$ must be apart, and indeed $c_2{*}$ lies above $a_2{*}$.

By making use of the abovementioned idea peculiar to production theory, we can now derive the *second version of decomposition equation.* Compared with the first version, this second one will turn out to be more analogous to the decomposition equation in consumption theory. For that purpose, Assumption (A6) needs to be a bit strengthened to the following assumption.

Assumption (A6′) For all $(w, p) \in W \times P$, $x\ (w, p)$ and $y\ (w, p)$ are *twice differentiable* in (w, p).

When Assumption (A6 ') is made, it is obvious by Lemmas 6.4 that $u\ (w, p)$ and $c\ (w, y)$ are also twice differentiable. We are now ready to derive the following theorem.

Theorem 6.8 (The Second Kind of Decomposition Equations) For all $(w, p) \in W \times P$, we have the following equations.

$$\partial x_i/\partial w_j = \partial u_i/\partial w_j - (\partial x_i/\partial p)\left(\partial p/\partial w_j\right) \text{ for all } i,j,$$

where $\partial p/\partial w_j \equiv [dp/dw_j]dw_i = 0\ (i \neq j),\ dy = 0$.

Proof We first note that under Assumption (A6′), $c\ (w, y)$ is now twice differentiable. Then, by help of Lemma 6.3 above, we find the following.

$$\partial c/\partial y = p. \tag{6.40}$$

Total differentiation of this equation yields the following.

$$\sum_i \left(\partial^2 c/\partial w_i \partial y\right) dw_i + \left(\partial^2 c/\partial y^2\right) dy = dp,$$

in which we have

$$dy = \sum_i \left(\partial y/\partial w_i\right) dw_i + \left(\partial y/\partial p\right) dp.$$

In particular, let $dw_i = 0$ for $i \neq j$, and $dy = 0$. Then, we obtain the following.

$$\left(\partial^2 c/\partial w_j \partial y\right) dw_j = dp, \tag{6.41}$$

$$(\partial y/\partial w_j)\, dw_j + (\partial y/\partial p)\, dp = 0. \tag{6.42}$$

If we differentiate Eq. (6.40) with respect to p only, we must have the following.

$$(\partial c^2/\partial y^2)\ (\partial y/\partial p) = 1,$$

which must imply the following.

$$(\partial y/\partial p) \neq 0.$$

Therefore, it is clearly seen by Eq. (6.42) that for an arbitrary change dw_j, a compensated change dp, making $dy = 0$, is uniquely determined as the following quantity.

$$dp = -dw_j\,(\partial y/\partial w_j)/(\partial y/\partial p). \tag{6.43}$$

If we take account of Eqs. (6.43) and (6.41), then we can thus derive the following equation.

$$\begin{aligned} \partial p/\partial w_j &\equiv [d\,p/d\,w_j]dwi = 0(i \neq j), dy = 0 \\ &= -(\partial\, y/\partial\, w_j)/(\partial\, y/\partial\, p) \\ &= \partial^2 c/\partial\, w_j\, \partial\, y. \end{aligned} \tag{6.44}$$

In the light of Theorems 6.6 (3) and 6.7 (1), we obtain the following.

$$\begin{aligned} \partial\, x_i/\partial\, w_j &= \partial\, x_j/\partial\, w_i \\ &= \partial\, u_j/\partial\, w_i + (\partial\, u_j/\partial y)(\partial\, y/\partial\, w_i). \end{aligned} \tag{6.45}$$

Note that by Theorem 6.5 (3), we have $\partial u_j/\partial w_i = \partial u_i/\partial w_j$. Moreover, taking advantage of Lemma 6.1 (2), Theorem 6.6 (4), and Eq. (6.44), we can derive the following.

$$\begin{aligned} (\partial\, u_j/\partial\, y)(\partial\, y/\partial\, w_i) &= -(\partial^2 c/\partial\, y\, \partial\, w_j)(\partial\, x_i/\partial\, p) - (\partial\, x_i/\partial\, p)(\partial^2 c/\partial\, w_j\, \partial\, y) \\ &= -(\partial\, x_i/\partial\, p)(\partial\, p/\partial\, w_j). \end{aligned}$$

Therefore, in the light of Eq. (6.45), we obtain the following.

$$\partial x_i/\partial w_j = \partial u_j/\partial w_i - (\partial x_i/\partial p)\ (\partial p/\partial w_j).$$

Thus, we have derived the desired result.
Q.E.D.

Theorem 6.8 gives us the *second version of decomposition equations in input demand theory.* The scale effect here is meant to represent the composite effect of the following two terms:

1. The change in output price so as to maintain the same output level as before, in spite of the change in the price of the input j.
2. The change in the input i corresponding to the variation in output price abovementioned.[11]

This second version of decomposition equations is as important as the first version, presumably being even more comparable to the famous Slutsky equations in consumption theory (see Slutsky (1915)). As an immediate result of this theorem, we can finally derive the following.

Theorem 6.9 (Property of the Expansion Effect) For all $(w, p) \in W \times P$, we have the following property.

$$EE_{ii} \equiv -(\partial x_i/\partial p)\,(\partial p/\partial w_i) \leqq 0, i = 1, \ldots, n. \tag{6.46}$$

Proof By making use of Eq. (6.44), Theorem 6.7 (2), and Lemma 6.1 (2), we can obtain the following sequence of equations.

$$\begin{aligned} -(\partial x_i/\partial p)(\partial p/\partial w_i) &= -(\partial x_i/\partial p)\left(\partial^2 c/\partial y \partial w_i\right) \\ &= -(\partial u_i/\partial y)(\partial y/\partial p)(\partial u_i/\partial y) = -(\partial y/\partial p)(\partial u_i/\partial y)^2, \end{aligned}$$

which should be non-positive by Theorem 6.6 (5).

Q.E.D.

The economic significance of this theorem is quite clear. It is recalled by Theorem 6.5 (4) that the following inequality holds.

$$SE_{ii} \equiv \partial u_i/\partial w_i \leqq 0. \tag{6.47}$$

In the light of Eqs. (6.46) and (6.47), we find that the substitution effect SE_{ij} and the scale effect EE_{ii} should always go in the same direction. Surely, this is also a remarkable contrast to consumption theory a la Hicks (1946), in which the substitution and income effects may go in opposite directions.

In short, input demand theory is input demand theory, thus being distinct from consumer demand theory. Although those two theories look somewhat similar, they

---

[11] By relying on the traditional calculus method, Hirota and Sakai (1969) succeeded in deriving the *second* version of decomposition equations. To our regret, however, their proof was rather sketchy and based on the Hessian matrix of the production function per se. We believe that a topological approach taken here is more general, and more elegant, than the previous method. This may look a small step, but surely a giant jump indeed.

are definitely different. We have to understand exactly how and to what extent they are analogous or distinct.

6.6 Some Remarks on the LeChatelier–Samuelson Principle

In this chapter, we have been manly concerned with the axiomatic foundations of input demand theory. While the approach taken here looks mathematical and rigorous, it has useful economic implications. In particular, it is noted that the total effect of a change in an input can be decomposed into the substitution and expansion effect. How and to what extent such decomposition in input demand is comparable to that in consumer demand is certainly a very important question to ask. To our surprise, such comparison has been rather neglected for long time in the economics literature. We do believe, however, that as the saying goes, it is better late than never.

We are ready here to do some economic applications and make final remarks. First, we note that our decomposition in input demand is closely related to the famous *LeChatelier–Samuelson principle.* Although this connection was pointed out by our friend Yoshihiko Otani (1982), we are going our own way to confirm it below.

Henri Louis LeChatelier (1850–1936) was a noted French scientist. He was best known for his work on his chemical equilibrium, which was to be called LeChatelier principle in the academia. In his classical work contained in Samuelson (1947, enlarged edition 1983), Samuelson boldly applied the principle to economic equilibrium, so that naming of the *LeChatelier–Samuelson principle* has been so popular in the academic world, especially in the economic profession (see Stiglitz (1966)).[12]

Seeing is believing! Samuelson (1947) once remarked as follows.

We have the following general theorem:

$$(dx_i/d\alpha_i)_0 \leqq (dx_i/d\alpha_i)_1 \leqq \cdots \leqq (dx_i/d\alpha_i)_{n-1} \leqq 0.$$

While the change in an x with respect to its own parameter is always negative regardless of the number of constraints, it is most negative when there are no constraints, only less so when there is a single constraint, and so forth, until the number of auxiliary constraints reaches the maximum possible, namely (n–1) (Samuelson 1947 & revised 1987, p. 38).

[12] According to Samuelson (1972), the LeChatelier Principle was discovered more than 100 years ago by LeChatelier, French chemist. Samuelson found, however, that it was a rather vaguely stated principle. So, he decided to make it more perfect and boldly applied to economic theory. This is the reason why this principle is now called the LeChatellier-Samuelson principle. Also see Dietzenbacher (1992) and Alexandrov and Bedre-Defolie (2017).

In our setting of input demand theory discussed so far, we have only to compare the case with no isoquant constraint and the one with a single constraint, namely the constraint that output y remains constant regardless a change in w_i. So, in terms of the input demand setting, we must have the following inequality.

$$(dx_i/dw_i)_0 \leqq (dx_i/dw_i)_1 \leqq 0. \tag{6.48}$$

If we rather want to follow the decomposition equation formula (6.45) abovementioned, then in the light of Eqs. (6.45), (6.46), and (6.47), we must find the following set of inequalities.

$$\partial x_i/\partial w_i = \partial u_i/\partial w_i - (\partial x_i/\partial p)\ (\partial p/\partial w_i),$$

or equivalently

$$TE_{ii} = SE_{ii} + EE_{ii}$$

Since the terms TE_{ii}, SE_{ii}, and EE_{ii} are all non-positive, we must have the following.

$$TE_{ii} \leqq SE_{ii} \leqq 0. \tag{6.49}$$

Needless to say, this equation clearly demonstrates the validity of the famous *LeChatelier –Samuelson principle* in input demand equilibrium. For this point, also refer to Rader (1968).

We now turn to the application of our input demand theory to some numerical examples. Hopefully, the following example will be helpful in understanding the fundamental difference between *the net effect and the gross effect* in input demand theory.

Let us consider the following production function of Cobb-Douglas type:

$$y = f\ (x_1, x_2, x_3) = x_1^{\alpha 1}\ x_2^{\alpha 2}\ x_3^{\alpha 3}, \alpha_1 + \alpha_2 + \alpha_3 < 0.$$

Let X and Y be the domain and range of f, respectively. If we assume that X is bounded above, so is Y, assuring assumption (A5). It can be proved that Assumptions (A1–A4) and (A6′) are also satisfied. It can be shown without difficulty that the matrix of the *total effect terms* is given as follows.

$$\begin{pmatrix} \partial x_1/\partial w_1 & \partial x_1/\partial w_2 & \partial x_1/\partial w_3 \\ \partial x_2/\partial w_1 & \partial x_2/\partial w_2 & \partial x_2/\partial w_3 \\ \partial x_3/\partial w_1 & \partial x_3/\partial w_2 & \partial x_3/\partial w_3 \end{pmatrix} = 1/[(1 - \alpha_1 - \alpha_2 - \alpha_3)py] \times G \tag{6.50}$$

where

$$G=\begin{pmatrix} -[(1-\alpha_2-\alpha_3)/\alpha_1]{x_1}^2 & -x_1x_2 & -x_1x_3 \\ -x_2x_1 & -[(1-\alpha_1-\alpha_3)/\alpha_2]{x_2}^2 & -x_2x_3 \\ -x_3x_1 & -x_3x_2 & -[(1-\alpha_1-\alpha_2)/\alpha_3]{x_3}^2 \end{pmatrix}$$

It is clearly seen from matrix G that any two inputs are *gross complements.*

Now, turning our attention to the matrix of *substitution effects*, we can easily derive the following.

$$\begin{pmatrix} \partial u_1/\partial w_1 & \partial u_1/\partial w_2 & \partial u_1/\partial w_3 \\ \partial u_2/\partial w_1 & \partial u_2/\partial w_2 & \partial u_2/\partial w_3 \\ \partial u_3/\partial w_1 & \partial u_3/\partial w_2 & \partial u_3/\partial w_3 \end{pmatrix} = 1/[(\alpha_1+\alpha_2+\alpha_3)py]\times H, \tag{6.51}$$

where

$$H=\begin{pmatrix} -[(\alpha_2+\alpha_3)/\alpha_1]{x_1}^2 & x_1x_2 & x_1x_3 \\ x_2x_1 & -[(\alpha_1+\alpha_3)/\alpha_2]{x_2}^2 & x_2x_3 \\ x_3x_1 & x_3x_2 & -[(\alpha_1+\alpha_2)/\alpha_3]{x_3}^2 \end{pmatrix}.$$

It is evidently seen that any two inputs are *net substitutes.*

Finally, in the light of Eqs. (6.50) and (6.51), we can show that the matrix of the *expansion effect terms* is provided as follows.

$$\begin{pmatrix} \partial x_1/\partial w_1 & \partial x_1/\partial w_2 & \partial x_1/\partial w_3 \\ \partial x_2/\partial w_1 & \partial x_2/\partial w_2 & \partial x_2/\partial w_3 \\ \partial x_3/\partial w_1 & \partial x_3/\partial w_2 & \partial x_3/\partial w_3 \end{pmatrix} - \begin{pmatrix} \partial u_1/\partial w_1 & \partial u_1/\partial w_2 & \partial u_1/\partial w_3 \\ \partial u_2/\partial w_1 & \partial u_2/\partial w_2 & \partial u_2/\partial w_3 \\ \partial u_3/\partial w_1 & \partial u_3/\partial w_2 & \partial u_3/\partial w_3 \end{pmatrix}$$
$$=1/[(\alpha_1+\alpha_2+\alpha_3)(1-\alpha_1-\alpha_2-\alpha_3)p\,y]\times J, \tag{6.52}$$

where

$$J=\begin{pmatrix} -x_1^2 & -x_1x_2 & -x_1x_3 \\ -x_1x_2 & -x_2^2 & -x_2x_3 \\ -x_1x_2 & -x_2x_3 & -x_3^2 \end{pmatrix}.$$

Let us compare the corresponding diagonal elements of the matrix G and the matrix J. Then, we will immediately find that the substitution and expansion effects of the price of an input on the demand of the same input must go in the same direction, thus intensifying each other. The famous the LeChatelier–Samuelson principle is surely valid as expected.

The name of Chemist LeChatelier is brilliantly shining not only in the chemical world but also in the economic world as well. Life may be short, but science is long and spreading indeed!

References

Alexandroff AD (1939) Almost everywhere existence of the second differential of a convex function and some properties of convex surfaces connected with it. Leningrad State University Annals, Leningrad, 6:3–35

Alexandrov A, Bedre-Defolie Ö (2017) LeChatelier-Samuelson principle in games and pass-through of shocks. J Econ Theory 168:44–54

Arrow KJ, Hahn FH (1971) General competitive analysis. Holden-Day, San Francisco

Basett LR, Borcheding TE (1969) 'Inferior factors' and theories of production and input demand: comment. Economica, New Series 38:321–322

Bear DVT (1965) Inferior inputs and the theory of the firm. J Polit Econ 73:287–289

Dietzenbacher E (1992) The LeChatelier-Samuelson principle revisited. J Econ 55:277–296

Diewert WE (1982) Duality approaches to microeconomic theory. Contained in Arrow KJ, Intriligator MD (eds) Handbook of mathematical economics, vol 2. North-Holland, pp 105–176

Diewert WE (2018) Duality in production. Vancouver School of Economics, Microeconomics working papers, no. 201

Fenchel W (1953) Convex cones, sets, and functions. Princeton University Press, Princeton

Ferguson CE (1966) Production, prices, and the theory of jointly-derived input demand functions. Economica, New Series 33:454–461

Ferguson CE (1968) 'Inferior factors' and the theories of production and input demand. Economica, New Series 35:140–150

Ferguson CE (1969) The neoclassical theory of production and distribution. Cambridge University Press, Cambridge

Friedman JM (1972) Duality principles in the theory of cost and production—revisited. Int Econ Rev 13:167–170

Hicks JR (1946) Value and capital, 2nd edn. Oxford University Press, Oxford

Hicks JR (1953–1954) A reply to Morshima. Rev Econ Stud 21:218–221

Hirota M, Sakai Y (1969) A note on net substitution and scale effects in production theory. Discussion paper, University of Rochester

Malinvaud E (1985) Lectures on microeconomic theory, Revised edition. North-Holland, Amsterdam

Mas-Colell A, Whinston MD, Green JR (1995) Microeconomic theory. Oxford University Press, Oxford

McFadden D (1966) Cost, revenue, and profit functions: a cursory review. University of California, Berkeley, Working paper no. 86

McKenzie LW (1957) Demand theory without a utility index. Rev Econ Stud 24:185–189

McKenzie LW (2002) Classical general economic theory. MIT Press, Cambridge

Mitra T, Nishimura K (eds) (2009) Equilibrium, trade, and growth: selected papers of Lionel W. McKenzie. MIT Press, Cambridge

Morishima M (1953a) A note on a point in value and capital. Rev Econ Stud 21:214–217

Morishima M (1953b) A rejoinder to Hicks. Rev Econ Stud 21:222

Nikaido H (1968) Convex structures and economic theory. Academic, London

Okuguchi K, Sakai Y, Ichioka O, Nagatani O (1989) Microeconomics (in Japanese). Yuhikaku Publishers, Tokyo

Otani Y (1982) A simple proof of the LeChatelier-Samuelson principle and the theory of cost and production. J Econ Theory 27:430–438

Rader T (1968) Normally, inputs are never gross substitutes. J Polit Econ 76:38–43
Rockafellar RT (1970) Convex analysis. Princeton University Press, Princeton
Sakai Y (1973) An axiomatic approach to input demand theory. Int Econ Rev 14:733–750
Sakai Y (1974) Substitution and expansion effects in production theory: the case of joint production. J Econ Theory 9:255–274
Sakai Y (1975) Substitution and expansion effects in the theory of the firm: new results in an old framework. Seikei Ronso, Hiroshima University 25(3):95–118
Samuelson PA (1947, enlarged ed. 1983) The foundations of economic analysis. Harvard University Press, Cambridge
Samuelson PA (1972) Maximum principles in analytical economics. Am Econ Rev 62:249–262
Scott RH (1962) 'Inferior' factors of production. Q J Econ 76:86–97
Shephard RW (1970) Theory of cost and production functions. Princeton University Press, Princeton
Slutsky E (1915) Sulla teoria del bilancio del consumatore. Gionale degli Economisti e Rivista di Statistica, 51: 1–26. Translated and contained in Stigler GJ, Boulding KE (eds) (1952) Readings in price theory, 27–56. Irwin, Illinois Press
Stiglitz J (ed) (1966) The collected scientific papers of Paul A. Samuelson, vol 1. MIT Press
Syrquin M (1970) A note on inferior inputs. Rev Econ Stud 37:591–598
Takagi T (1961) Principles of mathematical analysis, 3rd revised edition (in Japanese). Iwanaki Publishers, Tokyo
Uzawa H (1964) Duality principles in the theories of cost and production. Int Econ Rev 5:216–229
Varian HR (1999) Intermediate microeconomics: a modern approach. Norton, New York
Varian HR (2009) Microeconomic analysis. Norton, New York

Chapter 7
Substitution and Expansion Effects in Production: The General Case of Joint Production

Abstract This chapter aims to extend the axiomatic approach to input demand theory with a single output, which was taken in Chap. 6, to the more general case of *joint production.* Although such an extension from one output to several outputs seems to be straightforward, it nevertheless requires special care and careful interpretations. More specifically, it is concerned with a systematic examination of what happens when a firm that has been in equilibrium at certain prices of inputs and prices of outputs experiences a change in one of these prices. The *mutual effects between inputs between inputs* (*or between outputs*), the *cross effects between inputs and outputs* are newly explored by means of *various kinds of decomposition equations in production theory.* Those decomposition equations teach us that the *total effect of an input price (or an output price*) can be decomposed into the two separate effects; namely, the *substitution effect along on the old isoquant frontier (or the old transformation frontier)* and the *expansion effect along the new expansion path associated with input prices (or output prices).* Finally, a new definition of the *normal technology* is given to show that inputs are not gross substitutes, nor are outputs, and that the input–output relations are not r*egressive*.

Keywords Joint production · Comparative static analysis · Decomposition equations · Substitution and expansion effects · Normal technology · Regressive input–output relations

---

This chapter is a completely revised version of Sakai (1974), with many new figures. I am grateful to Edward Diewart (British Columbia), Michio Morishima (London), James Friedman (Rochester), Masayoshi Hirota (Rochester), Roger N. Folsom (Naval Postgraduate School), Yoshihiko Ohtani (Perdue), Trout Rader (Missouri), Edward Green (Pittsburgh), and Jack Ocks (Pittsburgh) for valuable comments and suggestions.

Y. Sakai, *Games, Decisions, and Markets*, New Frontiers in Regional Science: Asian Perspectives 74, https://doi.org/10.1007/978-981-99-5285-4_7

7.1 Mutual and Cross Effects in Production Theory

This chapter aims to extend the axiomatic approach to input demand theory with a *single output*, which was already adopted in Chap. 6, to the more general case of *joint production.* Although such an extension from only one output to several outputs seems to be rather straightforward, the history of economic theories has taught us that such extension work requires special care and cautious interpretations.[1]

More specifically, it is concerned with the question of what happens when the producer that has been in equilibrium at certain input prices and output prices experiences a change in one of those prices. Assume that the firm has been using certain input quantities and producing output quantities. Then, the interesting problem is in what way and to what extent these quantities are really affected.

In production theory, there are two different kinds of commodities inputs and outputs. We have to seek information about conditions governing the inputs demanded and the outputs supplied by the producer at given input–output prices to investigate the effects of price changes on the equilibrium of the firm. To this end, it is necessary to make a distinction between the following four sorts of effects.

1. **($W \rightarrow X$)** the mutual effect of a change in an input price on the demand for another input;
2. **($P \rightarrow Y$)** the mutual effect of a change in an output price on the supply of another output;
3. **($W \rightarrow Y$)** the cross effect of a change in an input price on the supply of an output
4. **($P \rightarrow X$)** the cross effect of a change in an output price on the demand for an input.

As may naturally be expected, the *mutual relations between inputs and between outputs* are studied through the following two effects: (1) **($W \rightarrow X$)** and (2) **($P \rightarrow Y$).** And, the *cross relations between inputs and outputs* are investigated via the following two effects: (3) **($W \rightarrow Y$)** and (4) **($P \rightarrow X$)**.

A modern development of input demand theory, following the impetus of Hicks (1946), Samuelson (1947), and Morishima (1953a, b), has confirmed that *the first effect (1) can be divided into two effects: a substitution effect along the old isoquant frontier and an expansion effect along the new expansion path associated with input prices.* Such decomposition in input demand theory has intensively been discussed by Hirota and Sakai (1969) and Sakai (1973, 1974, 1975). Although the term "net

[1] In the 1960s and the 1970s, I myself was a graduate student at the University of Rochester, and completing my Ph.D. thesis, an assistant professor at the University of Pittsburgh. In hindsight, at Rochester, I was so fortunate to be surrounded by many outstanding professors such as Lionel McKenzie, James Friedman, Edward Zabel, Ronald Jones, Emmanuel Drandakis, Akira Takayama, Hugh Rose, and Walter Oi. I would like to add that I was then so lucky to have many brilliant students including Jerry Green, Jose Scheinkman, Michihiro Ohyama, Masayoshi Hirota, and Richard Thaler. In hindsight, I can now say that this chapter is perhaps the product of the "Rochester spirits" in many possible ways. I am so grateful to many professors and fellow students for kind suggestions.

substitution and scale effects" was first employed in Hirota and Sakai (1969), a similar yet different expression "substitution and expansion effects" was fondly used in Sakai (1973, 1974, 1975) and will be used in our future work.

When we say goodbye to a very simple world of *single production* and dare to jump into a more general world of joint production, the decomposition aforementioned is expected to take a much more complicated and interesting form. *In fact, we can decompose the second effect (2) abovementioned into the following two separate effects;* namely, *a substitution effect along the old transformation frontier and an expansion* effect along the new scale path associated with out prices. It will be seen that the third effect (3) and the fourth effect (4) are also duly compounded effects.[2]

Interestingly enough, there is another long-standing problem concerning the possibility of gross complementarity among inputs and among outputs, together with the related problem of possibility of *regression in some of the input–output relations*. We are especially concerned with the *normal technology*. More exactly, we define the technology to be *normal* if one of the following conditions is met:

1. The marginal cost of an output tends to increase when the quantities of other outputs decrease or when the prices of inputs increase. (Simply, $MC_i\uparrow \Leftarrow y_j\downarrow$ or $w_s\uparrow$)
2. The marginal revenue of an input tends to go up when the quantities of other inputs go up or when the prices of outputs go up. (Simply, $MR_s\uparrow \Leftarrow x_t\uparrow$ or $p_i\uparrow$)

As will be seen below, if the technology is *normal*, then the inputs jointly employed by the firm are not gross substitutes, nor are the outputs jointly produced by it, and the relations between the inputs and the outputs are nor regressive. We believe that this is a very important result for understanding the consequences of the normal technology of the regular firm.[3]

The contents of this chapter will be outlined as follows. In Sect. 7.2, a system of axioms for the production structure is postulated. In terms of the structure, input and output possibility sets are exactly defined and their properties are carefully discussed with the help of illustrative figures. In Sect. 7.3, we successively introduce cost, revenue, and profit functions, with illustrative figures. Remarkably, the method by which the properties of these functions are obtained in this paper is closely related to the method of Lionel W. McKenzie (1957, 2002) used for consumer demand theory and also to the method of Uzawa (1964) and Shephard (1970) employed for duality between cost and production.

Section 7.4 deals with decomposition equations of the first kind in production theory. The total effect of price variation is split up into two separate effects, namely

---

[2]This and related points were well developed by Bear (1965), Basett and Borcheding (1969), Ferguson (1966, 1968, 1969), Syrquin (1970), Sakai (1973), and others. They are all concerned with the special case where there is only one output. In this chapter, we intend to go a step further by extending an axiomatic approach taken in Sakai (1973) to the very general case of joint production.

[3]The normal or typical situation is that an increased output price induces an increased input demand. If this is not the case, then we can say that the input–output relation is *regressive*. See Hicks (1946).

substitution and expansion effects. Section 7.5 is devoted to careful examinations of the properties of substitution, expansion, and total effects. It is observed that the substitution and expansion effects of a change in the price of an input (or an output) on the quantities of that input (or that output) should always go in the same direction. No doubt, this result confirms that the famous LeChatelier–Samuelson principle is nicely applicable in our production world. In Sect. 7.6, we attempt to obtain decomposition equations of the second kind. In so doing, we boldly introduce the idea of a *compensated change in the prices of outputs* (*or inputs*) when the price of a certain input (or output) varies. Finally, the concept of *normal technology* is introduced in Sect. 7.7. Then, it is shown that inputs are not gross substitutes, nor are outputs, and that the input–output relations are not regressive.

7.2 The Production Technology: Definitions and Properties

Let us consider a firm that uses n inputs and produce m outputs. We introduce the following notation for commodity bundles and price vectors:

$x = (x_1, ..., x_n)$, an input bundle.
$y = (y_1, ..., y_m)$, an output bundle.
$w = (w_1, ..., w_n)$, an input price vector.
$p = (p_1, ..., p_m)$, an output bundle.

It is assumed that x and y are non-negative vectors of real numbers, and that w and p are positive vectors of real numbers. We use the following notations:

$$W = \{w : w = (w_1, \ldots, w_n) > 0\}.$$

$$P = \{p : p = (p_1, \ldots, p_m) > 0\}.$$

The firm has a *production possibility set* T, namely the set of all input–output bundles $(-x, y)$, which are compatible with the technology available to it. For a series of discussions on the set T, see Arrow and Hahn (1971), Diewert (1982, 2018), Fenchel (1953), Malinvaud (1985), Mas-Colell et al. (1995)), McFadden (1966), Malinvaud (1985), McKenzie (2002), Mitra and Nishimura (2009), Nikaido (1968), Okuguchi et al. (Okuguchi et al. 1989), Rockafellar (1970), Samuelson (1947, revised ed. 1983), Stiglitz 1966, Samuelson (1972)), Shephard (1970), Takagi (1961), and Varian (1999, 2009).

Concerning the production possibility set T, we now postulate the following reasonable set of assumptions:

(T.1) T is nonempty; $(0, y) \in T$ implies $y = 0$.
(T.2) T is closed.
(T.3) T is bounded above.
(T.4) T is strictly convex.
(T.5) $(-x, y) \in T$ implies $(-x', y') \in T$ for any $x' \geqq x$ and for any $y' \leqq y$.

Assumption (T.1) asserts that zero inputs yield zero outputs. (T.2) states that for a sequence $\{(-x^k, y^k)\}$ ($k = 1, 2, \ldots$) of input–output bundles, if all the bundles $(-x^k, y^k)$ are feasible and if $(-x^k, y^k) \to (-x^0, y^0)$, then $(-x^0, y^0)$ is also feasible. As will be seen, the purpose of (T.3) is to force the profit function to be defined for any price vector (w, p). And, (T.4) means that if $(-x^0, y^0)$, $(-x^1, y^1) \in T$, $(-x^0, y^0) \neq (-x^1, y^1)$, $t \in (0, 1)$, and $(-x^t, y^t) = (1 - t)(-x^0, y^0) + t(-x^0, y^0)$, then we must have $(-x^t, y^t) \in \text{Int}\, T$, where Int T is the relative interior of T. Finally, (T.5) implies that T admits free disposal.[4]

Let us define a permissible input set X and a producible output set Y as follows.

$$X = \{x : x \geqq 0 \text{ and } (-x, y) \in T \text{ for some } y \geqq 0\}, \tag{7.1}$$

$$Y = \{y : x \geqq 0 \text{ and } (-x, y) \in T \text{ for some } x \geqq 0\}. \tag{7.2}$$

Under Assumptions (T.1)—(T.5), X and Y are nonempty and convex; X is bounded below, and Y is bounded above. Let us define the following sets.

$$X(y) = \{x \in X : (-x, y) \in T\}, y \in Y. \tag{7.3}$$

$$Y(x) = \{y \in Y : (-x, y) \in T\}, x \in X. \tag{7.4}$$

We say that $X(y)$ is an *input possibility set* associated with the attainable output y, namely the set of input bundles that can produce y. Likewise, $Y(x)$ is an *output possibility set* associated with the permissible input x, namely the set of output bundles that can be produced from x. Clearly, we find $X = \cup\{X(y): y \in Y\}$ and $Y = \cup\{Y(x): x \in X\}$.

Now, let us introduce the following:

$$I(y) = X(y) - \text{Int}\, X(y). \tag{7.5}$$

$$T(x) = Y(x) - \text{Int}\, Y(x). \tag{7.6}$$

Note that Int $X(y)$ and Int $Y(x)$ respectively denote the interior of $X(y)$ in X and the interior of $Y(x)$ in Y. Then, $I(y)$ is the familiar isoquant for a given y, and $T(x)$ is the well-known transformation frontier for a given x.

In the light of Assumptions (T.1)–(T.5), we can easily obtain the following properties of the input possibility set.

(X.1) For each $y \in Y$, then $X(y)$ is nonempty, closed, and strictly convex in X.

(X.2) If $x \in \text{Int}\, X(y)$, $y \in Y$, then $x \in X(y')$ for some $y' \geqq y$.

(X.3) If $x' \geqq x$, $y \in Y$, and $x \in X(y)$, then $x' \in X(y)$.

As for the output possibility set, we have readily the following properties.

---

[4] Note that Property (X.2) corresponds to Assumption (C.2) of Friedman (1972), which was introduced by him to show the continuity of $c(w, y)$ with respect to (w, y), not merely with respect to w. Our somewhat strong assumptions on T are required to derive the *differentiability* of $c(w, y)$ with respect w. Also see Uzawa (1964) and Shephard (1970).

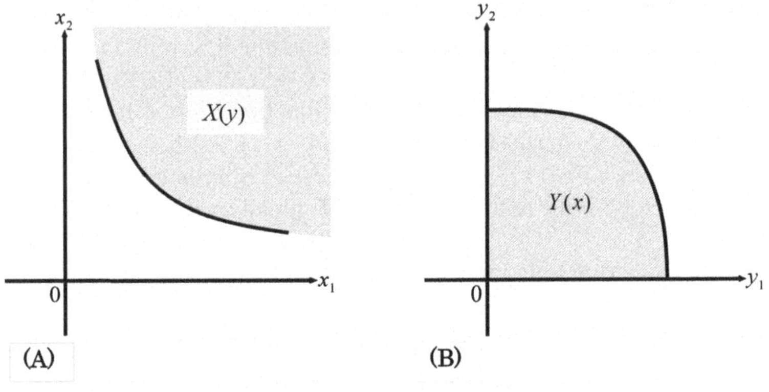

Fig. 7.1 Input possibility set $X(y)$ (**A**) and output possibility set $Y(x)$ (**B**)

(Y.1) For each $x \in X$, $Y(x)$ is nonempty, closed, and strictly convex in Y.
(Y.2) If $y \in \text{Int}\, Y(x)$, $x \in X$, then $y \in Y(x')$ for some $x' < x$ $(x' \geqq 0)$.
(Y.3) If $0 \leqq y' \leqq y$, $x \in X$, and $y \in Y(x)$, then $y' \in Y(x)$.

Hopefully, simple illustrations of the production possibility set T and related concepts would be very helpful in understanding such an axiomatic approach to production theory. As the saying goes, however, saying is one thing but doing is another. Here, we are concerned with the general model of joint production, with any inputs and many outputs. So, the simplest possible situation would be the situation under which the number of inputs and that of outputs are both limited to only two; therefore, $x = (x_1, x_2)$ and $y = (y_1, y_2)$. This implies that we need to draw a four-dimensional diagram for the set T, which seems to be a "mission impossible" unless some special devices are taken.

On the one hand, if we conveniently focus on two inputs x_1 and x_2, and one output y, then we can roughly draw an input possibility set $X(y)$ and an isoquant for a given y like Fig. 7.1(A). On the other hand, if we just focus on two outputs y_1 and y_2, and one input x, then we can have an output possibility set $Y(x)$, and a transformation frontier $T(x)$ for a given x like Fig. 7.1(B).

Fortunately, a very ingenious devise was once proposed by T. C. Koopmans (1964), a Nobel prize winner for economic science, in connection with the von Neumann model of general equilibrium and turnpike path. According to cool Koopmans, we need to cut a four-dimensional space down to three dimensions by means of normalization of inputs such that $x_1 + x_2 = 1$. Then, as is seen in Fig. 7.2, the production technology is represented by normalized points $z = (x_1, y_1, y_2)$, with each point z representing a normalized feasible activity $((x_1, 1-x_1), (y_1, y_2))$. Then, the normalized production set T is clearly convex and bounded above and also allows for free disposal. In short, it is a cleverly abridged three-dimensional version of the general production set X.

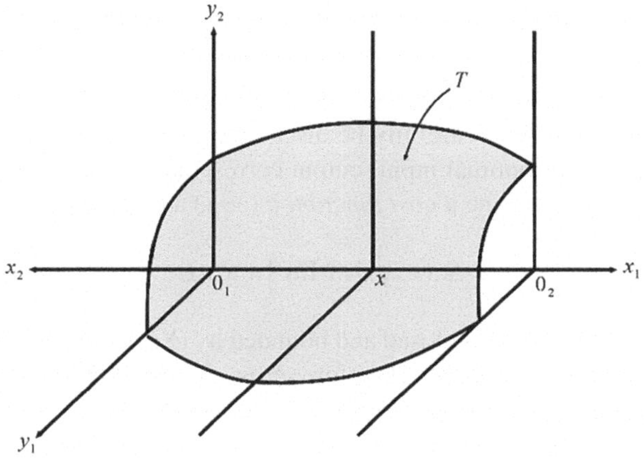

Fig. 7.2 The Koopmans diagram: an abridged three-dimensional version of the production possibility set T, which is closed, bounded above, and convex. Note that x_1 and x_2 are normalized such that $x_1 + x_2 = 1$

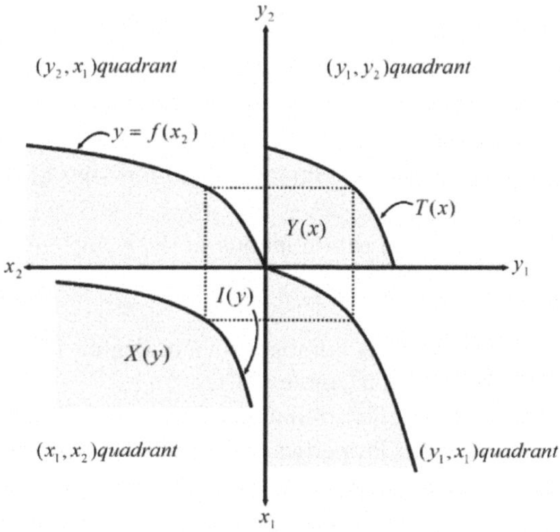

Fig. 7.3 Another version of the production possibility set T: limiting the set to its projections in four possible quadrants

Although the Koopmans diagram has its own advantage, we cannot deny the fact that the normalization process makes our production analysis here fairly artificial and slightly distorted. To avoid such unnecessary distortion, we attempt to find another useful device for T by limiting our attention to its projections in four possible quadrants; namely, (y_1, y_2) quadrant, (x_1, x_2) quadrant, (y_1, x_2) quadrant, and (x_1, y_2) quadrant. Take a careful look at Fig. 7.3. Then, in the first quadrant, we will find the output possibility set $Y(x)$ and the transformation frontier $T(x)$; and in the second quadrant, the projected production curve $y_2 = f(x_2)$; in the third quadrant, the input possibility set $X(y)$ and isoquant frontier $I(y)$; and in the fourth quadrant, the projected production curve $y_1 = f(x_1)$. Understandably, the four shadowed areas are all convex.

7.3 Cost, Revenue, and Profit Functions

In this section, cost, revenue, and profit functions will briefly be defined and their properties will carefully be discussed. In connection with these functions, compensated and optimal input–output correspondences will precisely be explored.

Let us define a *cost function* $c\ (w,\ y)$ as follows.

$$c\ (w, y) = \mathbf{Min}\ \{w\,x : x \in X\ (y)\}, (w, y) \in W \times Y. \tag{7.7}$$

Since $X\,(y)$ is closed and bounded by (X.1), the continuous function wx takes on a minimum on $X\ (y)$. Therefore, $c\ (w,\ y)$ is well defined. Then, we define a *compensated input correspondence* $u\ (w,\ y)$ as follows.

$$u(w, y) = \{x : x \in X(y) \text{ and } wx = c\ (w, y)\}, (w, y) \in W \times Y. \tag{7.8}$$

The correspondence $u(w,\ y)$ indicates the input vector demanded by the firm when the input price vector is w and the output bundle is to be y. Since $X\ (y)$ is strictly convex by (X.2), $u\ (w,\ y)$ must be the unique element of X minimizing $w\,x$ subject to $x \in X\ (y)$.

We are ready to summarize the properties of $c\ (w,\ y)$ and $u\ (w,\ y)$ as follows.

$$(\mathbf{C.1})\ c\ (w, y) \text{ is continuous in } (w, y), \text{differentiable in } w, \text{and } \partial c / \partial w_i = u_i\ (w, y), i = 1, \ldots, n. \tag{7.9}$$

(C.2) $c\ (w,\ y)$ is homogeneous of degree 1 and concave in w.
(C.3) $c\ (w,\ y)$ is convex in y.
(C.4) $u\ (w,\ y)$ is continuous in $(w,\ y)$ and homogeneous of degree 0 in w.

The proof of Properties of (C.1)–(C.4) is parallel to that of Lemma 6.1 in Chap. 6 above, which was given to the special case of single production and is safely omitted here; the only difference is that y is now a multi-valued vector. The cost function $c\ (w,\ y)$ and its properties are illustrated in Fig. 7.4.

Let us define a revenue function $r\ (p,\ x)$ and a compensated output correspondence $v\ (p,\ x)$ as follows.

$$r\ (p, x) = \mathbf{Max}\ \{p\,y : y \in Y\ (x)\}, (p, x) \in P \times X. \tag{7.10}$$

$$v\ (p, x) = \{y : y \in Y\ (x) \text{ and } p\,y = r\ (p, x)\}, (p, x) \in P \times X. \tag{7.11}$$

Since $Y\,(x)$ is closed and bounded from above, and strictly concave, $r\ (p,\ x)$ and $v(p,\ x)$ are both well defined. The correspondence $v\ (p,\ x)$ stands for the output bundle supplied by the firm when the output price is p and the input bundle is to be x. A symmetrical argument proves the following properties of $r\ (p,\ x)$ and $v\ (p,\ x)$.

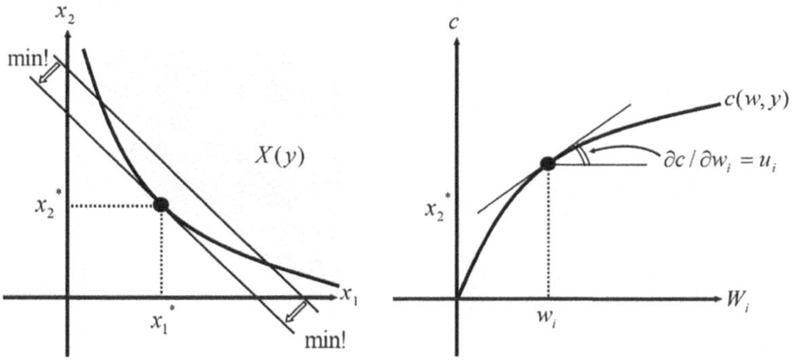

(A) $c(w, y) = w_1 x_1{}^* + w_2 x_2{}^*$ **(B)** $c(w, y)$ is concave in w

Fig. 7.4 The cost function $c\ (w, y)$ and its properties

(R.1) $r\ (p, x)$ is continuous in (p, x), differentiable in p, and $\partial r / \partial p_s = \mathrm{v}_s\ (p, x), s = 1, \ldots, n.$ (7.12)

(R.2) $r\ (p, x)$ is homogeneous of degree 1 and concave in p.
(R.3) $r\ (p, x)$ is convex in x.
(R.4) $v\ (p, x)$ is continuous in (p, x) and homogeneous of degree 0 in p.

When we compare Properties (C.1)–(C.4) and Properties (R.1)–(R.4), we can see that the cost structure and the revenue structure are nicely symmetrical. In particular, it is worth mentioning that, as is seen in (C.1), the rate of change of total cost when the price of an input changes is equal to the amount of compensated demand for the input. Symmetrically, as is seen in (R.1), the rate of change of total revenue when the price of an output changes is equal to the amount of compensated supply of the output. Symmetry is really beautiful and powerful! The revenue function and its properties are shown in Fig. 7.5.

Since the production possibility set T is closed and bounded by (T.2) and (T.4) above, the continuous function $py + w(-x)$ attains a maximum on T. Thus, we can define a *profit function* as follows.

$$\pi(w, p) = \textbf{Max}\ \{p\,y + w\,(-x) : (-x, y) \in T\}, (w, p) \in W \times P. \tag{7.13}$$

In terms of $\pi(w, p)$, we can define an *optimum input correspondence* $x\ (w, p)$ and optimum output correspondence $y\ (w, p)$ as follows.

$$(-x\ (w, p), y\ (w, p)) = \{(-x, y) \in T : p\,y - w\,x = \pi(w, p)\}. \tag{7.14}$$

Since T is strictly concave by (T.4), $(-x\ (w, p), y\ (w, p))$ is uniquely determined for each $(w, p) \in W \times P$. Surely, it is the unique combination of inputs and outputs that assures the maximum profit subject to the production technology.

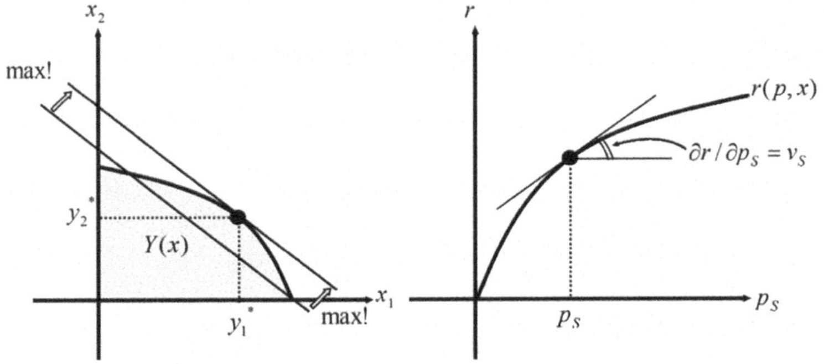

(A) $r(p, x) = p_s y_s^* + p_t y_t^*$

(B) $r(p, x)$ is concave in p

Fig. 7.5 The revenue function $r(p, x)$ and its properties

We are now ready to summarize the properties of $\pi(w, p)$, $x(w, p)$, and $y(w, p)$ as follows.

(P.1) $\pi(w, p)$ is differentiable in (w, p), and

$$\partial\pi/\partial w_i = -x_i(w, p), i = 1, \ldots, n. \tag{7.15}$$

$$\partial\pi/\partial p_s = y_s(w, p), s = 1, \ldots, n. \tag{7.16}$$

(P.2) $\pi(w, p)$ is homogeneous of degree 1 and convex in (w, p).

(P.3) $x(w, p)$ and $y(w, p)$ are both continuous and homogeneous of degree 0 in (w, p).

The proof of Properties (P.1)–(P.3) is quite analogous to that the proof of Lemma 6.2 in the last chapter and can thereby be omitted here, keeping in mind that y is now a multi-valued vector. Among those properties, (P.1) is undoubtedly the most important one and must be kept in mind. In plain English, it asserts that the rate of change of total profit when the price of an input changes is equal to the negation of the amount of demand for the input, whereas the rate of change of total profit when the price of an output price changes is equal to the amount of supply of the output. The profit function and its properties are well illustrated in Fig. 7.6.[5]

[5]Note that Property (P.1) was also proved by Arrow and Hahn (1971) in a slightly different way. It seems that different persons are going their own ways toward the same destination.

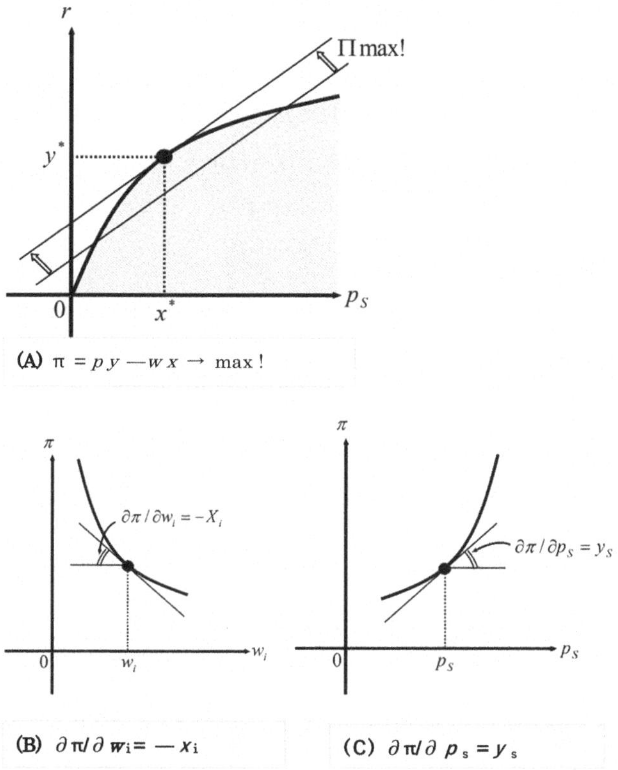

Fig. 7.6 Profit function and its properties

7.4 Decomposition Equations of the First Kind

In the last section, we sought information about conditions governing inputs demanded and output supplied by the firm at given input–output prices. The purpose of this section is to use them to discover how the inputs and outputs will change when these prices vary. In so doing, we will derive *decomposition equations of the first kind in production theory,* which are apparently analogous to the famous Slutsky equations in consumption theory (see Slutsky (1915)).

We first note the following important lemma, which will effectively be used for the derivation of decomposition equations in production theory.

Lemma 7.1 (Properties of Cost, Revenue, and Profit Functions) For any $(w, p) \in W \times P$, we obtain the following equations.

$$x\,(w,p) = u\,(w, y\,(w,p)). \tag{7.17}$$

$$c\ (w, y\ (w, p)) = w \cdot x\ (w, p). \tag{7.18}$$

$$y\ (w, p) = v\ (p, x\ (w, p)). \tag{7.19}$$

$$r\ (p, x\ (w, p)) = p \cdot y\ (w, p). \tag{7.20}$$

$$\begin{aligned} \pi(w, p) &= r(p, x(w, p)) - c(w, y(w, p)) \\ &= p \cdot y(w, p) - c(w, y(w, p)) \\ &= r(p, x(w, p)) - w \cdot x(w, p). \end{aligned} \tag{7.21}$$

Proof Since $x\ (w, p) \in X\ (y\ (w, p))$, we immediately find the following.

$$w \cdot x\ (w, p) \geqq\ c\ (w, y\ (w, p)) = w \cdot u\ (w, y\ (w, p)).$$

Assume $w{\cdot}x\ (w, p) > w{\cdot}u\ (w, y\ (w, p))$. Then, we have the following.

$$\pi(w, p) = p \cdot y\ (w, p) - w \cdot x\ (w, p) < p \cdot y\ (w, p) - u\ (w, y\ (w, p)),$$

which contradicts the maximum property of $\pi(w, p)$. Consequently, we must find the following.

$$w \cdot x\ (w, p) = c\ (w, y\ (w, p)) = w \cdot u\ (w, y\ (w, p)).$$

This implies $x\ (w, p) = u\ (w, y\ (w, p))$ because $X\ (y\ (w, p))$ is strictly convex by Assumption (X.1). Thus, Properties (1) and (2) are proved. The proof of Properties (3) and (4) proceeds in a similar way. Finally, since $\pi(w, p) = p{\cdot}y\ (w, p) - w{\cdot}x\ (w, p)$, we can obtain Property (5) by help of Properties (2) and (4).

Q.E.D.

The utmost importance of Lemma 7.1 cannot be exaggerated. As will be seen later, all kinds of decomposition equations in production theory, which surely constitutes the core of this chapter, will effectively be derived the lemma.

Let an input–output price vector (w, p) be given to the firm. Then, on the one hand, its profit-maximizing behavior uniquely determines the optimal input bundle $x\ (w, p)$ and the optimal output bundle $y\ (w, p)$, the corresponding quantity $c\ (w, y\ (w, p))$ stands for the total cost, which is to be paid by the firm for the production of $y\ (w, p)$. On the other hand, the quantity $r\ (p, x\ (w, p))$ indicates the total revenue, which is be obtained by the firm from the use of $x(w, p)$. And, the quantity $v\ (p, x\ (w, p))$ is the corresponding output bundle.

In order to facilitate our analysis, we now make an additional assumption:

(T.6) $x\ (w, p)$ and $y\ (w, p)$ are *twice differentiable* in (w, p).

When the new assumption (T.6) is newly introduced, it is seen from Lemma 7.1 that $\pi(w, p)$, $c\ (w, y)$, and $r\ (p, x)$ are also twice differentiable.[6]

We are now ready to establish and prove the following important theorem on decomposition equations in production theory.

Theorem 7.2 (The First Kind of Decomposition Equations) For all $(w, p) \in W \times P$, we have the following equations.

$$\partial x_i/\partial w_j = \partial u_i/\partial w_j + \sum\nolimits_t (\partial u_i/\partial y_t)\ (\partial y_t/\partial w_j) \text{ for any } i, j. \tag{7.22}$$

$$\partial x_i/\partial p_s = \sum\nolimits_t (\partial u_i/\partial y_t)\ (\partial y_t/\partial p_s) \text{ for any } i, s. \tag{7.23}$$

$$\partial y_s/\partial p_t = \partial v_s/\partial p_t + \sum\nolimits_i (\partial v_s/\partial x_j)\ (\partial x_j/\partial p_t) \text{ for any } s, t. \tag{7.24}$$

$$\partial y_s/\partial w_i = \sum\nolimits_j (\partial v_s/\partial x_j)\ (\partial x_j/\partial w_i) \text{ for any } s, i. \tag{7.25}$$

Proof From Lemma 7.1, we find the following equations.

$$x_i\ (w, p) = u_i\ (w, y\ (w, p)) \text{ for any } i,$$

$$y_s\ (w, p) = v_s\ (p_s, x\ (w, p)) \text{ for any } s.$$

Therefore, by applying a famous rule on the differentiation of a composite functions, we can immediately obtain the desired results.

Q.E.D.

As is seen in Eq. (7.22), Property (1) of Theorem 7.2 teaches us that *the total effect* of a *change in* w_i *on the demand on* x_i can be split into the following two effects: *a substitution effect* represented by the first term $(\partial u_i/\partial w_j)$, and *a scale effect* shown by the second summation term $\sum_t (\partial u_i/\partial y_t)\ (\partial y_t/\partial w_j)$.

In other words, the change in the price of an input j affects the demand for the amount of another input i in the following two different ways. First, it causes *the change in input price ratios, which in* turn *induces technical substitution among the* inputs along the old isoquant frontier. Second, it entails the changes in *all the outputs along* the *new expansion path associated with input prices,* thus contributing to the additional variation in the input i. When there is only one output, the second term is simply reduced to the term $(\partial u_i/\partial y_t)\ (\partial y_t/\partial w_j)$, which is indeed the same as the term that was obtained by Hirota and Sakai (1969), Syrquin (1970), and Sakai (1973,

[6]Under Assumptions (T.1)–(T.5) on the production technology, $\pi(w, p)$ is convex in (w, p), $c\ (w, y)$ is concave in w and convex in y, and $r\ (p, x)$ is convex in p and concave in x. Note here that convex (or concave) is really *almost everywhere twice differentiable* (see Alexandroff (1939)). Therefore, interestingly, Assumption (T.6) is not stringent as it appears. As will be seen in Theorem 7.2, the differentiability of $x\ (w, p)$ and $y\ (w, p)$ is sufficiently enough to obtain decomposition equations of the first kind. In short, there should be no substantial difference between convexity (or concavity) and twice differentiability.

1974, 1975)). Since there are generally m distinct outputs in the present case of joint production, however, the expansion effect term is written as the sum of m terms, each of which indicates the variation in the input i through the relevant output. As is shown in Eq. (7.23), Property (2) asserts that the cross effect of a change in p_s on x_i is also a compounded effect. That is, the change in the price of the output s entails the changes in all the outputs, each of which in turn contributes to the variation in the input i.

Likewise, it is seen by Eq. (7.24) representing Property (3) that the total effect of a change in p_t on y_s can be divided into the following two separate effects; namely, a substitution effect indicated by the first term $\partial v_s/\partial p_t$, and an expansion *effect* shown by the second summation term $\sum_i (\partial v_s/\partial x_j)(\partial x_j/\partial p_t)$.

First, the change in the price of the output t induces the change in output price ratios, which in turn entails technical substitution among the outputs along the old *transformation frontier*. Second, it also causes *the changes in all the inputs, along the* new expansion path, which in turn yield the subsequent change in the output s. More symbolically, we could say that $TE_{i\,j} = SE_{i\,j} + EE_{i\,j}$. Reflecting the fact that there are n distinct inputs, the expansion effect term can be expressed as the sum of n terms, each of which stands for the variation in the output s via the change in the relevant input. And finally, it follows from Eq. (7.25), representing Property (4), that the effect of a change in w_t is also a compounded effect of the following two effects. Namely, the effect in the price of the input i causes the changes in all the inputs, each of which in turn yields the variation in the output t. A series of decomposition equations in Theorem 7.2 play a key role in understanding the fundamental structure of production theory. Since Properties (1) and (3) are especially important, they are worthy of further examination.

For an illustrative purpose, let us consider the specific case of two inputs (x_1 and x_2) and a single output y. Let the price of the first input fall from w_1 to w'_1. Then, as Eq. (7.22) tells us, we will have the following pair of decomposition equations.

$$\partial x_1/\partial w_1 = \partial u_1/\partial w_1 + (\partial u_1/\partial y)(\partial y/\partial w_1):$$
$$(\textbf{total effect}) = (\textbf{substitution effect}) + (\textbf{expansion effect})$$

$$\partial x_2/\partial w_1 = \partial u_2/\partial w_1 + (\partial u_2/\partial y)(\partial y/\partial w_1):$$
$$(\textbf{total effect}) = (\textbf{substitution effect}) + (\textbf{expansion effect})$$

In Fig. 7.7, the movement from Point A to Point B and the movement from Point B to Point C respectively indicate the substitution effect and the expansion effect. Needless to say, the sum of these two effects shows the total effect.

Let us turn to the case of two outputs (y_1 and y_2) and a single input x. And now, let the price of the first output rise from p_1 to p'_1. Then, we will obtain the following pair of decomposition equations.

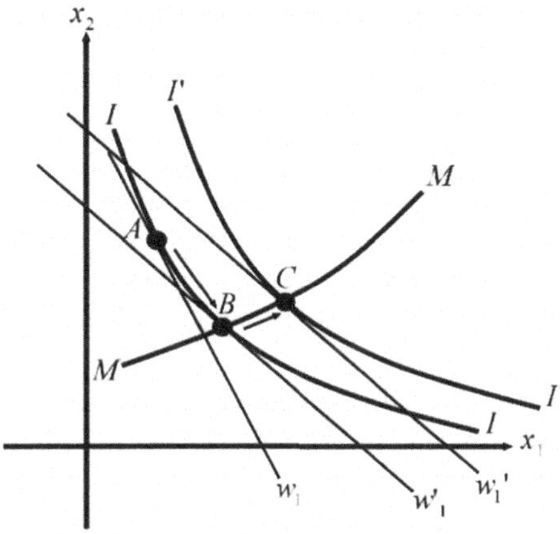

Fig. 7.7 The total effect ($A \rightarrow \rightarrow C$) = the substitution effect ($A \rightarrow B$) + the expansion effect ($B \rightarrow C$): a specific case of two inputs (x_1 and x_2) and a single output y

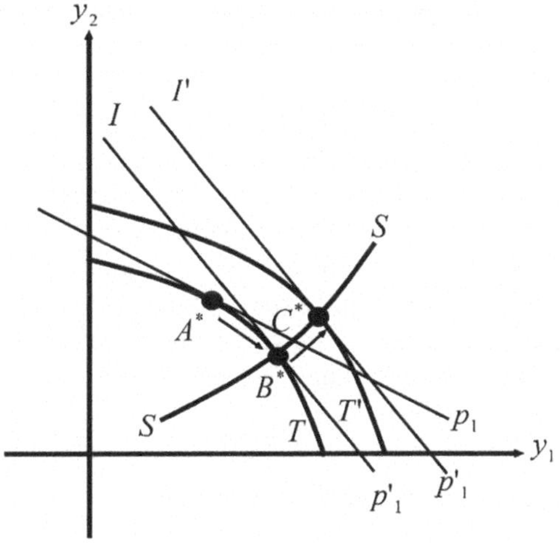

Fig. 7.8 The total effect ($A^* \rightarrow \rightarrow C^*$) = the substitution effect ($A^* \rightarrow B^*$) + the expansion effect ($B^* \rightarrow C^*$): another specific case of two outputs (y_1 and y_2) and a single input x

In Fig. 7.8, the movement from Point A^* to Point B^* and the movement from Point B^* to Point C^* respectively indicate the substitution effect and the expansion effect. The sum of these two effects clearly indicates the total effect.

$$\partial y_1/\partial p_1 = \partial v_1/\partial p_1 + (\partial v_1/\partial x)(\partial x/\partial p_1):$$

$$(\textbf{total effect}) = (\textbf{substitution effect}) + (\textbf{expansion effect})$$

$$\partial y_2/\partial p_1 = \partial v_2/\partial p_1 + (\partial v_2/\partial x)(\partial x/\partial p_1):$$

$$(\textbf{total effect}) = (\textbf{substitution effect}) + (\textbf{expansion effect})$$

7.5 Properties of the Substitution, Expansion, and Total Effects

The total effect of price variation, as we have seen above, is nicely decomposed into the two separately effects, namely the substitution and expansion effects. The properties of these three effects will carefully be investigated in this section.

To begin with, let us introduce some useful concepts, relating to the mutual relations between inputs and between outputs and the cross relations between inputs and outputs.

It is naturally expected that whereas a rise in the price of an input yields a decrease in the demand for the same input, a rise in the price of an output results in an increase in the supply of the same output. Keeping such reasonable relationship in mind, we may define net and gross substitutability/complementarity between inputs/outputs as follows.

On the one hand, the input i is called *a net substitute for the input j* if $\partial u_i/\partial w_j > 0$, and *a net complement for the input j* if $\partial u_i/\partial w_j < 0$. Likewise, the output s is said *a net substitute for the output t* if $\partial v_s/\partial p_t < 0$, and a *net complement for the output s substitute for the output t* if $\partial v_s/\partial p_t > 0$. On the other hand, the input i is called *a gross substitute for the input j* if $\partial x_i/\partial w_j > 0$, and *a gross complement for the input j* if $\partial x_i/\partial w_j < 0$. Likewise, the output s is said *a gross substitute for the output t* if $\partial y_s/\partial p_t < 0$, and a gross *complement for the output t* if $\partial y_s/\partial p_t > 0$.

Now, concerning the cross relationship between inputs and outputs, we may naturally expect that a rise in input price decreases an output supply. If such relation does not hold, we may say that it should be regressive, going against a normal course. More formally, we say that *the output s is regressive against the input i* if $\partial y_s/\partial w_t > 0$.[7]

We are ready to establish and prove a pair of the following interesting theorems.

Theorem 7.3 (Differential Properties of *u* (*w*, *y*)) For all $(w, y) \in W \times Y$, we have the following properties.

1. $\sum_j (\partial u_i/\partial w_j) \cdot w_j = 0$ for any i.
2. $\sum_i w_i \cdot (\partial u_i/\partial w_j) = 0$ for any j.
3. $\partial u_i/\partial w_j = \partial u_j/\partial w_i$ for any i, j.

[7] Alternatively, we may say the input i is *inferior with respect to the output s* if $\partial x_i \, \partial p_s < 0$, whereas the output s is *inferior with respect to the input i* if $\partial y_s/\partial w_i > 0$. Therefore, as far as production theory is concerned, "regression" and "inferiority" are virtually equivalent. See Hicks (1946).

4. The following matrix is *negative* semi-definite.

$$\begin{pmatrix} \dfrac{\partial u_1}{\partial w_1} & \cdots & \dfrac{\partial u_1}{\partial w_n} \\ \vdots & & \vdots \\ \dfrac{\partial u_n}{\partial w_1} & \cdots & \dfrac{\partial u_n}{\partial w_n} \end{pmatrix}.$$

Proof u (w, y) is homogeneous of degree zero in w by (C.4) above, so that application of the Euler theorem on a homogeneous function yields Property (1). To see Property (2), it is noted that the following inequality holds.

$$w \cdot u\,(w', y) \geqq \; w \cdot u\,(w, y) \text{ for any } w' \in W.$$

This means that the function $w{\cdot}u\,(w', y)$ of w' is minimized at $w' = w$. Therefore, the first-order condition for a minimum yields Property (2). To see Property (3), we note by means of (C.2) that the following equation holds.

$$\partial c/\partial w_i = u_i\;(w, y), i = 1, \ldots, n.$$

If we apply the Young theorem on the order of differentiation here, then we obtain the following.

$$\partial u_i/\partial w_j = \partial^2 c/\partial w_j \partial w_i = \partial^2 c/\partial w_i \partial w_j = \partial u_j/\partial w_i \text{ for } i, j = 1, \ldots, n.$$

Finally, since $c(w, y)$ is concave in w by (C.2) and Eq. (7.9) surely holds. Property (4) is easily derived.

Q.E.D.

In mathematics, symmetry is beautiful and powerful. Indeed, a symmetric argument will immediately lead to the following theorem.

Theorem 7.4 (Differential Properties of v (p, x)) For all $(p, x) \in P \times X$, we have the following properties.

1. $\sum_t (\partial v_s/\partial p_t) \cdot p_t = 0$ for any s.
2. $\sum_s p_s \cdot (\partial v_s/\partial p_t) = 0$ for any t.
3. $\partial v_s/\partial p_t = \partial v_t/\partial p_s$ for any s, t.

4. The following matrix is *positive* semi-definite.

$$\begin{pmatrix} \partial v_1/\partial p_1 & \cdots & \partial v_1/\partial p_m \\ \vdots & & \vdots \\ \partial v_m/\partial p_1 & \cdots & \partial v_m/\partial p_n \end{pmatrix}.$$

Undoubtedly, there appears outstanding symmetry between the input-related effect $\partial u_i/\partial w_j$ and the output-related effect $\partial v_s/\partial p_t$.

On the one hand, it follows from Property (4) of Theorem 7.3 that $\partial u_i/\partial w_i \leqq 0$ for any i, so that the slope of the compensated demand curve for any specific input tends to be negative. On the other hand, Property (4) of Theorem 7.4 implies that $\partial v_s/\partial p_s \geqq 0$ for any s, which demonstrates that the compensated supply curve for any specific output tends to be positive. These conclusions quite agree with common sense. According to Property (3) of Theorem 7.3 and Property (3) of Theorem 7.4, the substitution effects are mutually symmetrical between inputs and between outputs. Besides, it follows from Properties (1) and (2) of Theorem 7.3 and Properties (1) and (2) of Theorem 7.4 that there are limits on the amount of net complementarity possible among inputs and among outputs. In fact, whereas it is possible that all inputs (or outputs) are net substitutes for each other, it is not possible that they are all complements for each other.[8]

Theorem 7.5 (Differential Properties of *x* (*w*,*p*) and *y* (*w*,*p*)) For all $(w,p) \in W \times P$, we have the following properties.

1. $\sum_j (\partial x_i/\partial w_j) \cdot w_j + \sum_t (\partial x_i/\partial p_t) \cdot p_t = 0$ for any i.
2. $\sum_i (\partial y_s/\partial w_j) \cdot w_j + \sum_t (\partial y_s/\partial p_t) \cdot p_t = 0$ for any s.
3. $\partial x_i/\partial w_j = \partial x_j/\partial w_i$ for any i, j.
4. $\partial y_s/\partial p_t = \partial y_t/\partial p_s$ for any s, t.
5. $\partial x_i/\partial p_s + \partial y_s/\partial w_i = 0$ for any i, j.

6. The following matrix is *positive* semi-definite.

$$\begin{pmatrix} -\partial x_1/\partial w_j & -\partial x_1/\partial p_t \\ \partial y_s/\partial w_j & \partial y_s/\partial p_t \end{pmatrix}.$$

Proof It is noted that x (w, p) and y (w, p) are both homogeneous of degree 0 by (T.6) above. Therefore, application of the Euler theorem on homogeneous functions

[8] Thus, the substitution terms appear in both production theory and consumption theory and obey very similar laws.

yields Properties (1) and (2). And, from (P.1) above, we find the following equations:

$$\partial\pi/\partial w_i = -x_i \text{ for any } i,$$
$$\partial\pi/\partial p_s = y_s \text{ for any } s.$$

Hence, applying the Young theorem on the order of differentiation, we can derive the following series of equations:

$$\partial x_i/\partial w_j = -\partial^2\pi/\partial w_j\partial w_i = -\partial^2\pi/\partial w_i\partial w_j = \partial x_j/\partial w_i.$$
$$\partial y_s/\partial p_t = \partial^2\pi/\partial p_t\partial p_s = \partial^2\pi/\partial p_s\partial p_t = \partial y_s/\partial p_t.$$
$$\partial x_i/\partial p_s + \partial y_s/\partial w_i = -\partial^2\pi/\partial p_s\partial w_i + \partial^2\pi/\partial p_s\partial w_i = 0.$$

There, Properties (3), (4), and (5) are all assured.

Finally, since$\pi(w, p)$ is convex in (w, p) by (P.1) above, the following Hessian matrix on the left must be positive semi-definite.

$$\begin{pmatrix} \partial^2\pi/\partial w_j\partial w_i & \partial^2\pi/\partial p_t\partial w_i \\ \partial^2\pi/\partial w_j\partial p_s & \partial^2\pi/\partial p_t\partial p_s \end{pmatrix} = \begin{pmatrix} \partial x_i\partial w_j & -\partial x_i\partial p_t \\ \partial y_s\partial w_j & \partial y_s\partial p_t \end{pmatrix}.$$

This clearly demonstrates that Property (6) is assured.

Q.E.D.

If we look at Theorem 7.5 from an economic viewpoint, then we can derive many interesting implications.

First of all, it is observed as a special case of Property (6) that $x_i/\partial w_i \leqq 0$ for any i and $\partial y_s/\partial p_s \geqq 0$ for any s. Therefore, the demand curve for an input tends to be downward sloping, while the supply curve for an output tends to be upward sloping. Coupled with such observations, Properties (1) and (2) imply the following results. When a certain input (output) is gross complement for all other inputs (or all other outputs), it is not possible that it is regressive against all outputs (or all inputs). Such relationship between complementarity and regression has not been paid enough attention in the literature of microeconomics so far.

Properties (3) and (4) assert that the total effect is symmetrical between inputs and between outputs. In the light of Theorems 7.3 (3) and 7.4 (3), we already know that the expansion effect is also symmetrical between them. Here, we see a critical difference between the scale effect in production theory and the income effect in consumption theory, because the latter effect may not be symmetrical. Next, Property (5) implies that the input i is regressive against the output s if and only if the latter is against the former, thus showing nice symmetry of the cross relation between the input and the output. Finally, Property (6) also indicates the dominance of the own effects over the cross effects; indeed, for any i, j and any s, t, we find the following inequalities:

$$(\partial x_i/\partial w_i)\ (\partial x_j/\partial w_j) \geqq\ (\partial x_i/\partial w_j)\ (\partial x_j/\partial w_i)\ \text{for any}\ i, j.$$

$$(\partial y_s/\partial p_s)\ (\partial y_s/\partial p_s) \geqq\ (\partial y_s/\partial p_t)\ (\partial y_t/\partial p_s)\ \text{for any}\ s, t.$$

$$-(\partial x_i/\partial w_i)\ (\partial y_s/\partial p_s) \geqq\ -(\partial x_i/\partial p_s)\ (\partial y_s/\partial w_i)\ \text{for any}\ i, j.$$

We think that such predominance of the own effects may agree with common sense

We are now ready to establish and prove the following theorem.

Theorem 7.6 (Properties of the Expansion Effect) For all $(w, p) \in W \times P$, we have the following property.

1. $\sum_t (\partial u_i/\partial y_t)\ (\partial y_t/\partial w_i) \leqq 0$ for any i.
2. $\sum_j (\partial v_s/\partial x_j)\ (\partial x_j/\partial p_s) \leqq 0$ for any s.

Proof If we take Theorems 7.5 (5) and 7.2 (2) into account, then we find the following.

$$\begin{aligned}\sum\nolimits_t (\partial u_i/\partial y_t)(\partial y_t/\partial w_i) &= -\sum\nolimits_t (\partial x_i/\partial p_t)(\partial u_i/\partial y_t)\\ &= -\sum\nolimits_s \sum\nolimits_t (\partial u_i/\partial y_s)(\partial y_s/\partial p_t)(\partial u_i/\partial y_t),\end{aligned}$$

which is nonnegative by virtue of positive semi-definiteness of the matrix $(\partial y_s/\partial p_t)$ (see Theorem 5.6 (6)). This proves Property (1). Naturally, a symmetrical argument leads to Property (2).

Q.E.D.

Coupled with Theorems 7.3 and 7.4, the economic significance of Theorem 7.6 is now quite clear. It clearly assures that the substitution and scale effects always go in the same direction. In other word, the expansion effect cannot go in its own way, merely reinforcing the substitution effect.

In short, the possibility of the famous or infamous Giffen case is completely excluded in production theory. This is a remarkable contrast to traditional consumption theory, in which, as noticed by John Hicks (1946), the substitution and income effects may go in opposite directions. For this point, see Rader (1968) and Scott (1962).

As an important corollary of Theorem 7.6, we would like to discuss the question of how this theorem is related to the LeChatelier–Samuelson principle, which is one of the most famous principles in natural science, especially chemistry. In fact, we can straightforwardly establish and prove the following results.

Corollary 7.7 (The LeChatelier–Samuelson Principle in Production Theory)

1. $\partial x_i/\partial w_i \leqq \partial u_i/\partial w_i \leqq 0$ for any i. (Simply, $TE_{ii} \leqq SE_{ii} \leqq 0$ for any i.)

2. $\partial y_s/\partial p_s \leqq \partial v_s/\partial p_s \leqq 0$ for any s. (Simply, $TE_{ss} \geqq SE_{ss} \geqq 0$ for any s.)

Proof In the light of Theorems 7.2, if we put $i = j$, then we easily find the following inequality:

$$\partial x_i/\partial w_i = \partial u_i/\partial w_i + \sum_t (\partial u_i/\partial y_t)\,(\partial y_t/\partial w_i).$$

Besides, by virtue of Theorems 7.5 (6) and 7.6 (1), we find the following:

$$\partial u_i/\partial w_i \leqq 0 \text{ and} \sum_t (\partial u_i/\partial y_t)\,(\partial y_t/\partial w_i) \leqq 0.$$

Hence, we must obtain the following results:

$$\partial x_i/\partial w_i \leqq \partial u_i/\partial w_i \leqq 0 \text{ for any } i.$$

Clearly, this proves Property (1). The proof of Property (2) can proceed in a similar fashion.

Q.E.D.

The LeChatelier–Samuelson Principle was first applied to economic theory by Samuelson (1947, revised ed. 1983), and later developed by Otani (1982), Alexandrov and Bedre-Defolie (2017), and Dietzenbacher (1992).

Graphically, it is rather easy to represent the LeChatelier–Samuelson Principle in production theory. The principle simply says that Point C is in the eastern direction of Point B in Fig. 7.7 above, whereas Point C* is also in the eastern direction of Point B* in Fig. 7.8 above. If we attentively seek the truth, we can surely find the beauty and power of duality argument of production theory.

7.6 Decomposition Equations of the Second Kind

In this section, we will attempt to obtain decomposition equations of a different kind. In so doing, the idea of *compensated changes in output prices (or input prices) when the price of a certain input price (or output price) changes* will effectively be introduced.

Let us first remind ourselves of the way in which the famous decomposition equations in consumption theory, notably named the Slutsky equations, are derived (see Slutsky (1915)). When the price of a commodity varies, in consumption theory we have to consider a compensated change in a consumer's income so as to keep the preference or utility level as before the price change. Analogous to this, we intend to newly introduce *the idea of compensated changes in input prices (or output prices)* into production theory, which together with consumption theory firmly constitutes one of the keystones in the microeconomic structure. More exactly, when the price of an input changes, we image such compensated changes in the prices of all outputs as would induce the producer to maintain the same quantities of outputs as before the input price change; or when the price of an output varies, we suppose such compensated changes in the prices of all inputs as would the producer to keep the same quantities of inputs as before the output price change. It will be seen that the existence and uniqueness of such compensated changes in input prices or output prices are assured under Assumptions (T.1)–(T.6) above.

Introduction of the aforementioned idea into production theory will lead to decomposition equations of the second kind. To this end, we will first establish and prove the following lemma in a rigorous way.

Lemma 7.8 (Marginal Cost and Marginal Revenue) For all $(w, p) \in W \times P$, we obtain the following equations.

$$\partial c\left(w, y\left(w, p\right)\right)/\partial y_s = p_s \text{ for any } s. \tag{7.26}$$

$$\partial r\left(w, y\left(w, p\right)\right)/\partial x_i = w_i \text{ for any } i. \tag{7.27}$$

Proof From the definition of $\pi(w, p)$ and Lemma 7.1 (5), we find the following:

$$\pi(w, p) = p \cdot y\left(w, p\right) - c\left(w, y\left(w, p\right)\right) \geqq \ p\, y - c\left(w, y\right) \text{ for all } (w, p) \in W \times P.$$

Therefore, the first-order condition for a maximum yields the following equation.

$$\partial\left(p\, y - c\left(w, y\right)\right)/\partial y_s = 0 \text{ for any } s \text{ at } y = y\left(w, p\right).$$

Consequently, we can derive the following equation.

$$p\, \mathrm{s} = \partial c\left(w, y\left(w, p\right)\right)/\partial y_s \text{ for any } s.$$

This proves Property (1). The proof of Property (2) is parallel to the above. Q.E.D.

Lemma 7.8 demonstrates the profit-maximizing behavior of the producer from a different angle. It says that the marginal cost of an output is equal to the price of the output, whereas the marginal revenue of an input is equal to the price of the input. We are now in a position to establish one of the main results in this chapter in the following manner.

Theorem 7.9 (The Second Kind of Decomposition Equations) For all $(w, p) \in W \times P$, we have the following equations.

$$\partial x_i/\partial w_j = \partial u_i/\partial w_j - \sum\nolimits_t \left(\partial x_i/\partial p_t\right)\left(\partial p_t/\partial w_j\right) \text{ for all } i, j, \tag{7.28}$$

where $\partial p_t/\partial w_j \equiv [d\, p_t/d\, w_j] d\, wi = 0\ (i \neq j),\ d\, y_s = 0$.

$$\partial y_s/\partial p_t = \partial v_s/\partial p_t - \sum\nolimits_j \left(\partial y_s/\partial w_j\right)\left(\partial w_j/\partial p_t\right) \text{ for all } s, t, \tag{7.29}$$

where $\partial w_j/\partial p_t \equiv [d\, w_j/d\, p_t]dp_s = 0\ (s \neq t),\ d\, x_i = 0$.

Proof In the light of Eq. (7.26) in Lemma 7.8 (1) above, we note the following:

$$\partial c\left(w, y\left(w, p\right)\right)/\partial y_t = p_t \text{ for any } t. \tag{7.30}$$

Total differentiation of this equation yields the following.

$$\sum\nolimits_i \left(\partial^2 c/\partial w_i \partial y_t\right) d\, w_i + \sum\nolimits_s \left(\partial^2 c/\partial y_s\, y_t\right) d\, y_s = d\, p_t \text{ for any } t, \tag{7.31}$$

where we note the following:

$$d\, y_s = \sum\nolimits_{\mathrm{i}} \left(\partial y_s/\partial w_i\right) d\, w_i + \sum\nolimits_{\mathrm{t}} \left(\partial y_s/\partial p_t\right) d\, p_t \text{ for any s.} \tag{7.32}$$

Now, let us especially put $d\, w_i = 0$ for any $i \neq j$, and $d\, y_s = 0$ for any s. Then, Eq. (7.32) is simplified to the following:

$$\sum\nolimits_t \left(\partial y_s/\partial p_t\right) d\, p_t = -\left(\partial y_s/\partial w_j\right) d\, w_j,$$

which can be rewritten in matrix notation as follows.

$$\begin{pmatrix} \partial y_1/\partial p_t & \cdots & \partial y_1/\partial p_t \\ \vdots & & \vdots \\ \partial y_m/\partial p_t & \cdots & \partial y_m/\partial p_m \end{pmatrix} = -dw_j \begin{pmatrix} \partial y_1/\partial w_j \\ \vdots \\ \partial y_m/\partial w_j \end{pmatrix}. \tag{7.33}$$

Then, it is clearly seen by Eq. (7.33) that, for an arbitrary change dw_j in the input price j, a vector of compensated changes $(d\, p_1, \ldots, d\, p_m)$ in all output prices is uniquely determined. Indeed, to this end, let us define the following:

$\partial p_t/\partial w_j \equiv [\mathrm{d}\, p_t / \mathrm{d}\, w_j]\mathrm{d}w_i = 0\ (i \neq j)$, $\mathrm{d}\, y_s = 0$ for any i.

Then, from Eq. (7.33), we obtain the following equation in matrix notation, thus uniquely determining the vector of partial derivatives $(\partial p_1/\partial w_j, \ldots, \partial p_m/\partial w_j)$:

$$\begin{pmatrix} \partial p_1/\partial w_j \\ \vdots \\ \partial p_m/\partial w_j \end{pmatrix} = -\begin{pmatrix} \partial y_1/\partial p_1 & \cdots & \partial y_1/\partial p_m \\ \vdots & & \vdots \\ \partial y_m/\partial p_m & \cdots & \partial y_1/\partial p_m \end{pmatrix}^{-1} \begin{pmatrix} \partial y_1/\partial w_j \\ \vdots \\ \partial y_m/\partial w_j \end{pmatrix}. \tag{7.34}$$

Now, letting $dw_i = 0$ for any $i \neq j$, and $d\, y_s = 0$ for any s, and also taking account of Eq. (7.31), we find the following:

$$\partial p_t/\partial w_j = \left(\partial^2 c/\partial w_i \partial y_t\right) \text{ for any } t. \tag{7.35}$$

In view of Theorem 7.5 (3) and Theorem 7.2 (1), we find the following:

$$\begin{aligned}\partial x_i/\partial w_j &= \partial x_j/\partial w_i \\ &= \partial u_j/\partial w_i + \sum\nolimits_t (\partial u_j/\partial y_t)(\partial y_t/\partial w_i).\end{aligned}$$

From Theorem 7.3 (3), we note $\partial u_j/\partial w_i = \partial u_i/\partial w_j$. Besides, taking account of (C.1) above and Theorem 7.5 (5) together with Eq. (7.35), we find the following:

$$\begin{aligned}(\partial u_j/\partial y_t)(\partial y_t/\partial w_i) &= \left(\partial^2 c/\partial w_i \partial y_t\right)(-\partial x_i/\partial p_t) \\ &= -(\partial x_i/\partial p_i)(\partial p_t/\partial w_j).\end{aligned}$$

Consequently, we find the following equation:

$$\partial x_i/\partial w_j = \partial u_i/\partial w_j - \sum\nolimits_t (\partial x_i/\partial p_t)\ (\partial p_t/\partial w_j),$$

which assures Property (1).

To see Property (2), we first note the following equation.

$$\partial r\ (p, x\ (w, p))/\partial x_j = w_j \text{ for any } j. \tag{7.36}$$

Total differentiation of Eq. (7.36) yields the following:

$$\sum\nolimits_s \left(\partial^2 r/\partial p_s \partial x_j\right) dp_s + \sum\nolimits_i \left(\partial^2 r/\partial x_i\, x_j\right) dx_i = dw_j \text{ for any } j.$$

where we note the following equation:

$$dx_i = \sum\nolimits_i (\partial x_i/\partial w_i)\ dw_i + \sum\nolimits_t (\partial x_i/\partial p_t)\ d\,p_t \text{ for any } i.$$

Then, on the same line as the above, it is seen that, for an arbitrary change dp_t in the output price t, a vector of compensated changes (dw_1, ..., dw_n) in all input prices is uniquely determined, and indeed we find the following:

$$\begin{aligned}\partial w_j/\partial p_t &\equiv \left[dw_j/dp_t\right] dp_s = 0 (s \neq t),\ \ dx_i = 0 \\ &= \partial^2 r/\partial p_t \partial x_j \ \text{ for any } t.\end{aligned}$$

With the above remark in mind, the proof of Property (2) proceeds in a similar way.

Q.E.D.

Theorem 7.9 provides us with different, perhaps more interesting, versions of the expansion effects. In fact, the expansion effect of the change in the price of the input j on the demand for the input i is now expressed as *the sum of m terms such that the* t-th *term* ($t = 1, ..., m$) *represents the composite effect of the following two terms:*

1. the change in the price of the output t so as to maintain the same output as before the input price changes.
2. the change in the input i corresponding to the variation in output abovementioned.

Similar interpretation may be given to the scale effect of the change in the price of the output t: namely, it is written as *the sum of n terms such that the j* th term ($j = 1, ..., n$) *indicates the composite effect of the following two terms:*

1. the change in the price of the input j so as to keep the same inputs as before the output price change.
2. the change in the output s responding to the variation in input price aforementioned.

To sum up, we have derived two different types of decomposition equations in production theory. Roughly speaking, the total effect can be divided into the substitution and scale effects. While the producer's scale effect and the consumer's income effect seem to be similar, they are not the same. On the one hand, the income effect plays a role of a trouble maker in consumption theory, sometimes going in the opposite direction from the substitution effect and producing the infamous Giffen paradox. On the other hand, the expansion effect acts as a sympathizer, always going the same direction and producing no paradox at all.

In the long history of microeconomics, the existence of the scale effect in production effect has long been neglected. We would strongly believe that such unfortunate neglect should immediately be corrected. As the saying goes, it is surely better late than never.

7.7 The Normal Case: A Generalization of Trout Rader's Analysis

In the above, we have systematically investigated the question of what happens when a firm that have been in equilibrium at certain input–output prices experiences a change in one of these prices. We have especially been concerned with deriving various kinds of decomposition equations in production theory, which have been unfortunately neglected in the vast literature on the theory of the firm. We hope that our results obtained so far will be helpful in promoting microeconomic theory.

Once in his nice paper, our good friend Trout Rader (1968) claimed that "normally, inputs are never gross substitutes." More specifically, for a single-valued production function $y = f(x_1, ..., x_n)$, he wisely defined that the technology is *normal* if the following conditions are met.

$$(\textbf{TR})\ \partial^2 f/\partial x_i \partial x_j \geqq\ 0 \text{ for any } i, j.$$

The condition (TR), named after our old friend Trout Rader, requires that all the inputs be "cooperative" in the production process in the sense that an increase in an input tends to increase the marginal product of another input. Rader succeeded to show that under condition (TR), all inputs tend to be gross complements, which mathematically means $\partial x_i/\partial w_j \leqq 0$. We want to extend the Rader result to the more general case of joint production. On the basis of the shapes of the cost and revenue curves, we say that the technology is *normal* if the following set of conditions (YS1)–(YS4) are all satisfied.[9]

$$(\textbf{YS1})\ \partial^2 c/\partial y_t \partial y_s \leqq\ 0 \text{ for any } s \neq t.$$

$$(\textbf{YS2})\ \partial^2 c/\partial w_i \partial y_s \geqq\ 0 \text{ for any } i, s.$$

$$(\textbf{YS3})\ \partial^2 r/\partial x_j \partial x_i \geqq\ 0 \text{ for any } i \neq j.$$

$$(\textbf{YS4})\ \partial^2 r/\partial p_s \partial x_i \geqq\ 0 \text{ for any } i, s.$$

In the normal case, on the one hand, the marginal cost of an output tends to increase when the quantities of other output *decrease* or when the prices of inputs increase. On the other hand, the marginal revenue of an input tends to go up when the quantities of other inputs go up, or when the prices of outputs go up.

Theorem 7.10 (The Normal Case) If the technology is normal, then we have the following results.

1. $\partial x_i/\partial w_j \leqq 0$ for any i, j.
2. $\partial y_t/\partial p_s \geqq\ 0$ for any s, t.
3. $\partial x_i/\partial p_s \geqq\ 0$ for any i, s.
4. $\partial y_s/\partial w_i \leqq 0$ for any i, s.

Proof In the light of Lemma 7.8 (2), we find the following equation in vector notation.

$$(\partial r/\partial x_i) = (w_i). \tag{7.37}$$

Differentiation of Eq. (7.37) yields the following equation in matrix notation.

[9] Our definition of the "normal case" here can be regarded as a natural extension of Trout Rader's definition to the more general case of joint production. As soon as Sakai met Rader at the University of Rochester, they became very close friends. Fortunately, a few years later, they met again at the University of Pittsburgh, with Jack Ocks being their nice go-between.

$$\left(\partial^2 r/\partial x_i \partial x_j\right)\left(\partial x_i/\partial w_j\right) = I_n. \tag{7.38}$$

where I_n denotes the identity matrix of order n. By virtue of Eq. (7.36), we immediately find the following.

$$\left(\partial x_i/\partial w_j\right) = \left(\partial^2 r/\partial x_i \partial x_j\right)^{-1}. \tag{7.39}$$

Since $r(w, p)$ is concave in x by (R.3) above, the matrix $(\partial^2 r/\partial x_i \partial x_j)$ is negative semi-definite. Moreover, its off-diagonal elements are all non-negative when the technology is normal (see (YS3) above). Therefore, by applying the well-known Mosak theorem on matrix inversion, its inverse must be a non-positive matrix, which assures Property (1).

Similarly, from Lemma 7.8 (1), we can derive the following equation in matrix notation.

$$\left(\partial y_s/\partial p_t\right) = \left(\partial^2 c/\partial y_s \partial y_t\right)^{-1}. \tag{7.40}$$

The matrix $(\partial^2 c/\partial y_s \partial y_t)$ is positive semi-definite by (c.3) above, and its off-diagonal elements are all non-positive by assumption (see Condition (YS1) above), so that its inverse must be a non-negative matrix by the Mosak theorem. This proves Property (2).

Next, in the light of Theorem 7.2 (2) and (C.1) above, we find the following.

$$\begin{aligned}\partial x_i/\partial p_s &= \sum\nolimits_t (\partial u_i/\partial y_t)(\partial y_t/\partial p_s) \\ &= \sum\nolimits_t \left(\partial^2 c/\partial w_i \partial y_t\right)(\partial y_t/\partial p_s),\end{aligned}$$

which is non-negative by Assumption (YS2) and Property (2) above.

Finally, from Theorem 7.5 (5), we obtain the following.

$$\partial y_s/\partial w_i = -\,\partial x_i/\partial p_s.$$

This is clearly non-positive by Property (3).

Q.E.D.

Theorem 7.10 teaches us what the normal case of production technology is all about. In fact, it gives us the rules that govern the market conduct of the producer. In the normal situation, as Trout Rader (1968) has shown, the inputs employed by the producer are not gross substitutes. Not only that, normally, the outputs produced are not gross substitutes either, and the input–output relations should not be regressive at all.[10]

In conclusion, we can say that production theory is as beautiful and powerful as it should be.

References

Alexandroff AD (1939) Almost everywhere existence of the second differential of a convex function and some properties of convex surfaces connected with it. Leningrad State University Annals, Leningrad, 6:3–35

Alexandrov A, Bedre-Defolie Ö (2017) LeChatelier-Samuelson principle in games and pass-through of shocks. J Econ Theory 168:44–54

Arrow KJ, Hahn FH (1971) General competitive analysis. Holden-Day, San Francisco

Basett LR, Borcheding TE (1969) 'Inferior factors' and theories of production and input demand: comment. Economica, New Series 38:321–322

Bear DVT (1965) Inferior inputs and the theory of the firm. J Polit Econ 73:287–289

Dietzenbacher E (1992) The LeChatelier-Samuelson principle revisited. J Econ 55:277–296

Diewert WE (1982) Duality approaches to microeconomic theory. Contained in Arrow KJ, Intriligator MD (eds) Handbook of mathematical economics, vol. II. North-Holland, pp 105–176

Diewert WE (2018) Duality in production. Microeconomics working paper, no. 2, Vancouver School of Economics

Fenchel W (1953) Convex cones, sets, and functions. Department of Mathematics, Princeton University

Ferguson CE (1966) Production, prices, and the theory of jointly-derived input demand functions. Economica, New Series 33:454–461

Ferguson CE (1968) 'Inferior factors' and the theories of production and input demand. Economica, New Series 35:140–150

Ferguson CE (1969) The neoclassical theory of production and distribution. Cambridge University Press, Cambridge

Hicks J (1946) Value and capital, 2nd edn. Oxford University Press, Oxford

Hirota M, Sakai Y (1969) A note on net substitution and scale effects in production theory. Discussion paper, University of Rochester

Koopmans TC (1964) Economic growth at a maximal rate. Q J Econ 78:355–394

Malinvaud E (1985) Lectures on microeconomic theory. Revised edition. North-Holland

Mas-Colell A, Whinston MD, Green JR (1995) Microeconomic theory. Oxford University Press, Oxford

McFadden D (1966) Cost, revenue, and profit functions: a cursory review. Working paper no. 86, Institute of Business and Economic Research, University of California, Berkeley

McKenzie LW (1957) Demand theory without a utility index. Rev Econ Stud 24:185–189

McKenzie LW (2002) Classical general economic theory. MIT Press, Cambridge

Mitra T, Nishimura K (eds) (2009) Equilibrium, trade, and growth: selected papers of Lionel W. McKenzie. MIT Press, Cambridge

Morishima M (1953a) A note on a point in value and capital. Rev Econ Stud 21:214–217

Morishima M (1953b) A rejoinder to Hicks. Rev Econ Stud 21:222

[10]These results were ambiguously anticipated, but not rigorously established, by Hicks (1946). Also see Morishima (1953a, b). We are very happy to see that the Hicks-Morishima-Sakai connection emerges here.

Nikaido H (1968) Convex structures and economic theory. Academic, London
Okuguchi K, Sakai Y, Ichioka O, Nagatani O (1989) Microeconomics (in Japanese). Yuhikaku Publishers, Tokyo
Otani Y (1982) A simple proof of the LeChatelier-Samuelson principle and the theory of cost and production. J Econ Theory 27:430–438
Rader T (1968) Normally, inputs are never gross substitutes. J Polit Econ 76:38–43
Rockafellar RT (1970) Convex analysis. Princeton University Press, Princeton
Sakai Y (1973) An axiomatic approach to input demand theory. Int Econ Rev 14:733–750
Sakai Y (1974) Substitution and expansion effects in production effects in production theory: the case of joint production. J Econ Theory 9(3):255–274
Sakai Y (1975) Substitution and expansion effects in the theory of the firm: new results in an old framework. Seikei Ronso, Hiroshima University 25(3):95–118
Samuelson PA (1947, revised ed. 1983) The foundations of economic analysis. Harvard University Press
Samuelson PA (1972) Maximum principles in analytical economics. Am Econ Rev 62:249–262
Scott RH (1962) 'Inferior' factors of production. Q J Econ 76:86–97
Shephard RW (1970) Theory of cost and production functions. Princeton University Press, Princeton
Slutsky E (1915) Sulla teoria del bilancio del consumatore. Gionale degli Economisti e Rivista di Statistica, 51: 1–26. Translated in Stigler GJ, Boulding KE (eds) (1952) Readings in price theory: 27–56. Irwin, Homewood
Stiglitz J (ed) (1966) The collected scientific papers of Paul A. Samuelson, vol 1. MIT Press
Syrquin M (1970) A note on inferior inputs. Rev Econ Stud 37:591–598
Takagi T (1961) Principles of mathematical analysis, 3rd revised edition (in Japanese). Iwanami Publishers, Tokyo
Uzawa H (1964) Duality principles in the theories of cost and production. Int Econ Rev 5:216–229
Varian HR (1999) Intermediate microeconomics: a modern approach. Norton, New York
Varian HR (2009) Microeconomic analysis. Norton, New York

Part III
Markets

Chapter 8
Interdependence of Several Markets: The Hicks–Morishima Approach Reconsidered

Abstract This chapter aims to shed some new light on the Hicks–Morishima approach to the interdependence of several markets. In spite of its rather simple and ambitious framework for the interdependence of several markets, it is quite unfortunate that this approach has been rather neglected in the academic circle. We suppose that there are several reasons for this. First, the traditional general equilibrium approach developed by Lionel W. McKenjie, Gerald Debreu, and Kenneth W. Arrow exclusively works with the good space rather than the price space. In contrast, the Hicks–Morishima approach based on Hicks' classical book *Value and Capital* exclusively operates on the price space, thus against the current mainstream of economic theory. Next, the majority of economics readers are usually familiar with the straightforward notion of demand and supply curves, but not with the twisted concept of excess demand curves. It is one of my main purposes to mend such unfortunate tendency, presumably proceeding toward the establishment of a new grand system of social science. We can learn new lessons from old teachings.

Keywords John Hicks · Michio Morishima · *Value and Capital* · Excess demand curves · General equilibrium analysis · Comparative statics

---

This chapter is a completely revised version of Sakai (2010). I am grateful to John Hicks (Oxford), Michio Morishima (London), Masayoshi Hirota (Tokyo Keizai), Tetsuya Nosse (Kobe Commercial), Lionel McKenzie (Rochester), James Friedman (Rochester), Peter Nijkamp (Free Univ.), Asatoshi Maeshiro (Pittsburgh), Arnold Katz (Pittsburgh), Toshikazu Ito (Ryukyu), Tatsuo Hatta (Osaka), Koji Okuguchi (Tokyo Metropolitan), and Yoshiro Higano (Tsukuba) for valuable suggestions.

Y. Sakai, *Games, Decisions, and Markets*, New Frontiers in Regional Science: Asian Perspectives 74, https://doi.org/10.1007/978-981-99-5285-4_8

8.1 Michio Morishima and John Hicks: The Two Superstars of Economics

There are the two economics superstars who have greatly influenced my academic career. They are Michio Morishima, Osaka University and London School of Economics, and John R. Hicks, London School of Economics and Oxford University. To tell the truth, both of them already passed away. Academically, however, they seem to be still alive: indeed, their outstanding accomplishments will never be forgotten in economic science. The main aim of this chapter is to reconsider and shed new light on a topic named the Hicks–Morishima approach to the interdependence of several markets. Although such approach has been rather neglected for some time, we are sure that even now, it is worthy of serious reconsideration, presumably giving a new guide toward an integrated grand theory of social science in the twenty-first century. No doubt, this should be the direction of future research to be strongly desired by both Morishima and Hicks in their late years.

When Sakai was a graduate student at the University of Rochester, he was fortunate to meet with Mr. Masayoshi Hirota, a "very serious man with highly samurai spirits," who so fondly argued night and day with me about life philosophy, world matters, and of course economic science. According to Hirota, we should not have any fears about the seemingly high levels of American universities. Osaka University seemed to be at least on a par with, and even better than, University of Rochester. Hirota proposed Sakai to do a joint work on general equilibrium theory with the final aim of publishing academic papers in professional journals of high prestige. We worked very hard by utilizing every space of the Rochester campus, namely the university library, class rooms, student dining rooms, and the graduate student center. After some time of our joint research, there came our final product, Hirota and Sakai (1969), which aimed to open a new horizon in theoretical studies in production theory. Although our joint paper had a humble yet attractive title "On Substitution and Scale Effects in Production Theory," it was quite unfortunate that because of some minor technical reasons, it missed a publication opportunity.

In hindsight, I am sure that my joint research experience with Mr. Hirota, or "Osaka samurai scholar," has constantly given us a great stimulus for the execution of our research project on economic theory. Indeed, having been encouraged a great deal by the Hirota–Sakai cooperation, I myself continued to learn many advanced mathematical tools, which were applicable not only to production theory but also to consumption theory. The final result of my continuous research effort was the completion of my doctoral dissertation, namely Sakai (1972) *Axiomatic Foundations of Consumption and Production Theories*, submitted to The University of Rochester, with my main thesis advisor being Professor Lionel W. McKenzie. After successfully finishing my Ph.D. work and publishing many theoretical papers in international journals, I could spend a very happy life at Pittsburgh, in which Professor

Asatoshi Maeshiro and his wife Kazuko kindly helped me in many ways. All those nice things happened to my rather long American life in the period 1968–1976.[1]

I can vividly recall that at the very end of 1973, my important meeting with Michio Morishima took place in New York City. More exactly, Morishima was invited to deliver a special lecture at Econometric Society North American Meeting at New York Hilton Hotel, with Professor Lawrence W. Klein of Pennsylvania serving as a respectable chairman. The audience filled the hall to overflowing. Professor Akira Takayama, a noted international economist, sat in the very front seat, taking so many pictures. All the people seemed to have a very happy time by enjoying the presence of the memorable trio, namely Morishima, Klein, and Takayama. Then, Morishima stood up and delivered a historical speech: "Kark Marx is so great! I have no doubt that he is academically still alive after one hundred years of his death!" In an instant, the whole audience stood up and applauded his powerful speech.[2]

In what follows, I would like to just pick up *Value and Capital*, the great book of John Hicks (1939, 2nd ed. 1946), carefully discussing how closely Hicks and Morishima have been connected by this masterpiece. A famous writer Naoya Komuro (2004) once remarked:

> During the Second World War, Morishima served as a naval cadet whose duty was to break a secret code of the enemy. When he left home toward a battlefront, he carried an important book with himself. That book was nothing but a newly published text *Value and Capital* by Hicks, then a rising star economist of the United Kingdom, one of Japan's hostile countries Nevertheless, Morishima, who faithfully conducted his secret mission of code breaking, dared to exert all his energy to fully understand the Hicks new book, and safely could return home (Komoro 2004).

In his later years, Morishima (1999a, b) got a sudden surge of nostalgia for his youth:

> I [Morishima] was studying economics at Kyoto University just after the Second World War was over. My favorite book was John Hicks' great book *Value and Capital.* One day, just like a bolt from the blue, the following sentence got into my eyes: "Let the excess demand

[1]In September 1972, I submitted my Ph.D. dissertation to the University of Rochester, with Professors Lionel W. McKenzie and James Friedman being my thesis advisers. As was seen in Sakai (1972), the thesis consisted of the two parts. The first part carefully discussed axiomatic approaches to consumption theory, the second part mathematical foundations of production theory. I would like to point out that Chap. 3 in Part II of this book, titled *An Axiomatic Approach to Input Demand Theory,* was originally based on my joint work with Dr. Masayoshi Hirota with further refinements. Very fortunately, a still further revision of this chapter was published in *International Economic Review,* which was the prestigious academic journal jointly established by Michio Morishima of Osaka University and Lawrence Klein of the University of Pennsylvania.

[2]More exactly speaking, Professor Morishima's invited lecture was titled *Marx in the Light of Modern Economic Theory.* It was nothing but a Walrus memorial lecture at the North American Meeting of Econometric Society. Although it was first read at Hilton Hotel, New York, in December 28, 1973, it was later published as a leading article of *Econometrica* (see Morishima 1974). It was really lucky that I myself was right there as an attentive audience. This constitutes one of never-to-be-forgotten memories in my long academic life.

for one good increase. Then, not only will its price will rise, but also the prices of other substitutive goods will rise, whereas the prices of complementary goods will decline." Although I was quite impressed by Hicks' remark, I nevertheless noticed that he carelessly neglected the undeniable truth that any economic activity including market trading should operate at some distance in a space. My question which naturally arose in my mind was how and to what extent Hicks' laws of market aforementioned would have to change by the introduction of complication factors including distance and space. In retrospect, I can now recall that in my student days, I was literally swamped with such very difficult problems. (Morishima 1999a, b)

It was quite fair to say that in those days, Morishima stood high and alone in the Japanese academia: in fact, the research level of Morishima went far beyond the one of any ordinary Japanese economist. He was a man of courage and determination. The major aim of this chapter is to reconsider Hicks's law of market exchange in a new modern fashion, thus further extending the Hicks–Morishima approach with my own flavor attached to it.

I myself have met with Hicks several times. My most memorable moment took place when I attended the International Conference in Honor of J.R. Hicks, at Bologna, Italy, in 1991. It gave me a great and unforgettable memory. Besides, Professor Hirofumi Uzawa and his lovely wife Hiroko were attending such a honorable conference, in fact serving as a co-chairman together with a noted Italian economist.

Now, it is high time to outline the contents of this chapter. The second section will deal with a simple model of general equilibrium the excess demand approach a la Hicks and Morishima. The third session will discuss many non-normal cases. The issues of instability, multiple equilibrium, and nonexistence will be successively taken up. In the fourth section, we will turn to the problem of comparative statics. The impact of parameter changes on equilibrium values will carefully be investigated. Final remarks will be made in the fifth section, with reference to Morishima's life-long dream of a grand synthetic approach to a so-called "symphonic economics."

8.2 A Simple Model of General Equilibrium: The Excess Function Approach a la John Hicks and Michio Morishima

Saying is one thing. But doing is another. Now let us discuss the working and performance of a simple model of general equilibrium. To this end, we will adopt the excess demand approach a la Hicks and Morishima. The Hicks–Morishima approach is a very effective one to shed light on interrelation of excess demand curves in the price space for an investigation of the working and performance of general equilibrium. Although we focus on a simple two-good model in this chapter, its generalization to any several good models would require no difficult task.

Honestly speaking, in spite of its rather simple and ambitious framework, it is quite unfortunate that the Hicks–Morishima approach has been long neglected in the academic circle. We suppose that there are several reasons for this. First of all, the traditional general equilibrium approach, developed by L. W. McKenzie (1954, 1959), G. Debreu (1959), and K. W. Arrow & G. Debreu (1954), Arrow and Hahn (1971) and others, exclusively works with the good space rather than the price space. In contrast, the Hicks–Morishima approach exclusively operates on the price space, thus against the current mainstream of economic theory. Next, the majority of economics readers are usually familiar with the straightforward notion of demand and supply curves, but not with the twisted concept of excess demand curves. And of course, many people prefer simplicity to complexity.

We suppose that Morishima would apparently agree with those reasons. In order for the excess demand approach to become very familiar to the academic circle, we would have to do a sort of extraordinary work for publicity and sales effort. Morishima had special courage to write a new style of economics text, Morishima (1984) *The Economics of Industrial Society*, in which he adopted the Hicks–Morishima approach again. Unfortunately, history repeated itself: It again received a rather dull response. Since Sakai regards himself as an "academic son" of Morishima and also as an "academic grandson" of Hicks, he should be very strong against the neglected stream of academia, eagerly arguing usefulness and applicability of the Hicks–Morishima Approach.

8.2.1 A Pair of Two Goods, Which Are Either Substitutive or Complementary

In reality, there seem to be many diversified markets, which are more or less interlocked with each other. If we take a simple example from a female clothes market, it is true that skirt and jeans markets are mutually interrelated. Their relations, however, are much more complicated than a man in the street would usually imagine. The demand for female clothes may be influenced by so many possible factors. First of all, a young stylish girl may want to be dressed in the height of fashion: indeed, being behind the fashion would be the most terrible thing to do. If wearing slim jeans, but not wide skirt, becomes more fashionable than before, then there would be so many young girls who want to change their fashion style from skirt to jeans. Besides, if skirt becomes more expensive due to an increased cost of fabric, then there would be a tendency that the sale of jeans as a good substitute of skirt will rise. Moreover, the overall demand for clothes, jeans or skirt, is determined by the purchasing power of buyers. When the economic conditions become improved (or worsened), people want to spend more (or less) money for clothing, so that the sales of skirt and jeans will overall tend to go up (or down).

This may not be all of the story. Let us recall that we live in the world of risk and uncertainty. If we consider people's forecasts and expectations, the demand and supply of a good should get much more complicated.

For instance, even if the price of a skirt has risen "this week," no one can exactly tell what will happen to that price "next week." If a young girl expects it to rise higher next week, she has a good incentive to buy it this week. Besides, even if her income stays sluggish this month, yet she expects it to go up significantly the next month, she might feel rich and affluent even now, possibly showing positive buying behavior. Similar reasoning will be applicable to the supply side, not only the demand side. In general, if the trader expects the sale to be sluggish this year, but to be increasing next year, he is likely to positively respond to his optimistic forecast.

To sum up, human beings are complicated, and so are markets. In fact, a skirt market and a jeans market are so intricate that they cannot simply be analyzed. It is the Hicks–Morishima approach that can shed a nice analytical light on the entangled interdependence of several markets. In historical perspective, that approach was originated in Hicks (1939, 2nd ed. 1946, Chapter 5) and later highly developed in Morishima (1984) in many possible ways. To our regret, however, time seems to pass too slowly to recognize a masterpiece: indeed, the magnificent Hicks–Morishima approach has been long undervalued or neglected. As we have said repeatedly, it is our sincere desire to revive the approach once more and hopefully attach some new findings and flavors to it.

8.2.2 A Simple Two-Good Model: The Starting Point

At the starting point of discussion, let us begin our market model analysis with a simple case of two goods, x_1 and x_2. More specifically, let us suppose that x_1 stands for tea and x_2 coffee. Further assume that the price of x_1 and the one of x_2 are respectively denoted by p_1 and p_2.

It is natural to assume that the amount of demand for x_1 is dependent not only on p_1 but also on p_2. Since tea and coffee are presumably in competitive relations, they are formally regarded as "substitutive" in microeconomic theory. Therefore, the demand function, D_1, for good 1 may be formulated as follows:

$$\boldsymbol{D_1} : x_1 = -p_1 + 0.5p_2 + 3. \tag{8.1}$$

It is supposed here that there are two different routes by which the amount of demand of coffee, x_1, rises. The first route is the *direct* one by means of a fall in p_1, the price of coffee per se. This may be called the "own effect" of a price change. The second route is the *indirect* one in the sense that the price rise of tea as a substitute of coffee causes the change of demand from tea to coffee. This may be named the "cross effect" of a price change. As is clearly seen in Eq. (8.1), it is supposed that the first own effect overpowers the second cross effect.

Concerning the supply side, we assume that its direction of change is exactly the opposite of the demand side. More specifically, let us consider the following linear supply function:

$$\boldsymbol{S_1} : x_1 = p_1 - 0.5\, p_2 - 1. \tag{8.2}$$

Let the price of coffee rise. Then, we obtain the following two effects. First, the distributor intends to increase the supply of coffee in the market. This is the "own effect" of a more expensive coffee. Second, when the price of tea falls, the distributor not only wishes to sell tea less, but also he tends to sell coffee more. This is the "cross effect" or the "spillover" effect over the two markets. Besides, it is supposed here that the first own effect overwhelms the second cross one.

We are now in a position to introduce the concept of an excess demand function. Note that excess demand means "demand minus supply": namely, $E_1 = D_1 - S_1$. Thus, in the light of Eqs. (8.1) and (8.2), we can easily derive the excess demand function for x_1 as follows:

$$\boldsymbol{E_1 = D_1 - S_1} : x_1 = -2p_1 + p_2 + 4. \tag{8.3}$$

This demonstrates that the amount of excess demand for good 1 is *inversely* proportional to its price (the own effect), whereas the former amount is in *direct* proportion to the price of good 2 (the cross effect). Besides, understandably, the own effect overpowers the cross one.

Market equilibrium of good 1 is realized when its demand and supply are just equal, namely its excess demand completely vanishes. It is at this equilibrium point that the price of good 1 ceases to change. On the one hand, if the demand for good 1 exceeds its supply, then the excess demand occurs, implying that the price of good 1 must go up. On the other, if the demand is less than the supply, then the excess supply occurs, so that the price has to fall. This is a famous adjustment process a la Leon Walras, which may be summarized as follows:

$$\boldsymbol{E_1 > 0} : D_1 > S_1 \text{ (disequilibrium)} \rightarrow p_1 \text{ goes up.}$$

$$\boldsymbol{E_1 = 0} : D_1 = S_1 \text{ (equilibrium)} \rightarrow p_1 \text{ does not change.}$$

$$\boldsymbol{E_1 < 0} : D_1 < S_1 \text{ (disequilibrium)} \rightarrow p_1 \text{ goes down.}$$

Let us take a look at Fig. 8.1. In Chart (A), the straight line on the price plane represents the excess demand function: $-2p_1 + p_2 + 4 = 0$. (Note that this function can also be rewritten as $p_2 = 2p_1 - 4$.) Since the two goods are substitutes, the line is positively sloped. Moreover, reflecting the fact that the own effect overpowers the cross effect, the slope of the straight line for p_1-axis should be steeper than the one for p_2-axis, so that the former slope must exceed 45°.

Suppose that a pair of prices (p_1, p_2) lies right on the straight line E_1. Then, p_1 does not change at all and maintains the same value as before. Clearly, the excess demand line $E_1 = 0$ divides the whole price plane into two areas. In the *left* area

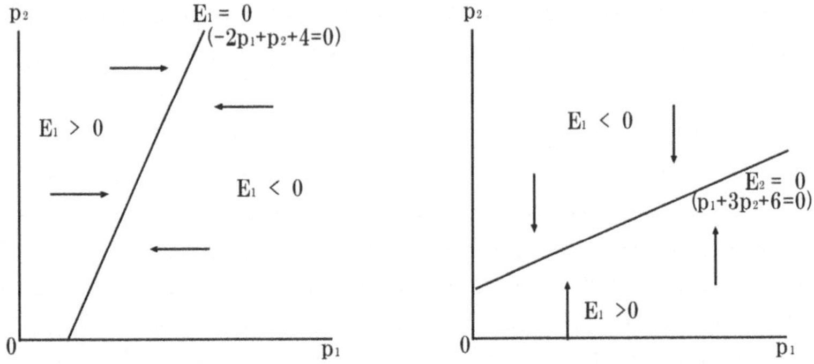

(A) Excess demand for x1: changes in p1 and p2

(B) Excess demand for x2: changes in p1 and p2

Fig. 8.1 The Relationship between the Two Markets

where $E_1 > 0$, p_1 rises in a direction indicated by the right-directed arrow (→). In contrast, the demand line $E_1 = 0$ divides the whole price plane into two areas. In the *left area* where $E_1 > 0$, p_1 rises in a direction indicated by the right-directed arrow (→). In contrast, in the *right area* where $E_1 < 0$, p_1 falls in an opposite direction indicated by the left-directed arrow (←).

A similar argument will apply for good 2 or coffee. So, we can formulate the demand, the supply, and the excess demand functions in the following way:

$$\boldsymbol{D_2} : x_2 = 0.5\,p_1 - 2p_2 + 4. \tag{8.4}$$

$$\boldsymbol{S_2} : x_2 = -0.5\,p_1 + p_2 - 2. \tag{8.5}$$

$$\boldsymbol{E_2} = \boldsymbol{D_2} - \boldsymbol{S_2} : x_2 = p_1 - 3p_2 + 6. \tag{8.6}$$

This teaches us that the amount of excess demand for good 2 is inversely proportional to its price p_2 (the own effect), whereas the former amount is in direct proportion to p_1, the price of good 1 (the cross effect). Besides, as before, the own effect overpowers the cross one.

The Walras-type adjustment process for good 2 may be formulated analogously:

$$\boldsymbol{E_2 > 0} : \boldsymbol{D}_2 > \boldsymbol{S}_2 \text{ (disequilibrium)} \rightarrow p_2 \text{ rises.}$$

$$\boldsymbol{E_2 = 0} : \boldsymbol{D}_2 = \boldsymbol{S}_2 \text{ (equilibrium)} \rightarrow p_2 \text{ does not change.}$$

$$\boldsymbol{E_2 < 0} : \boldsymbol{D}_2 < \boldsymbol{S}_2 \text{ (disequilibrium)} \rightarrow p_2 \text{ falls.}$$

Let us take a look at Fig. 8.1 again. In Chart (B), the straight line $E_2 = 0$, or $p_1 - 3p_2 + 6 = 0$, represents the excess demand for good 2. Note that this line may be rewritten as $p_1 = 3p_2$ 6. This is the case in which the two goods are substitutes and each own effect is dominating, so that the excess demand line is positively sloped and its "slope for p_2-axis" must exceed 45 degrees. The directions of price adjustment p_2 are indicated by arrows in the figure: namely, p_2 rises in the lower area, but falls in the upper area.

8.2.3 General Equilibrium of Two Markets: The Simplest Case

We are ready to get into the world of "general equilibrium" with the two goods, in which the demand and supply of each good are just equal and the changes of prices stop moving. In other words, the general equilibrium position looks like a stationary state.

Now, let us carefully investigate what happens if the two chart diagrams (A) and (B) of Fig. 8.1 as well as the corresponding sets of arrows are superimposed. Then, we will immediately see that a very charming float diagram like Fig. 8.2 appears. At the intersection point Q of the two excess demand lines, $E_1 = 0$ and $E_2 = 0$, the following two equations must be satisfied:

$$-2p_1 + p_2 + 4 = 0. \tag{8.7}$$

$$p_1 - 3p_2 + 6 = 0. \tag{8.8}$$

Solving Eqs. (8.7) and (8.8) for p_1 and p_2 simultaneously, we can see that a pair of coordinates of the intersection point Q is shown by $Q = (18/5, 16/5)$. Therefore, the equilibrium prices of p_1 and p_2 are respectively 18/5 (= 3.6) and 16/5 (= 3.2).[3]

To sum up, the intersection point Q indicates the point of "general equilibrium" of the two markets. Remarkably, it should be a stable point as well. Note that any price movements may be shown by the arrows of the pair (p_1, p_2). In whatever direction the price pair may move in Fig. 8.2, it is eventually destined to converge at the equilibrium point Q.

---

[3] A group of chart diagrams like Fig. 8.2 were first introduced by Hicks (1939). Morishima, who admired Hicks so much, intended to make the chart diagrams revived, with the aim of establishing his own "new economics." To our regret, however, such revival of the Hicks–Morishima approach has not been very successful so far. My sincere intention here is to give such an ingenious approach one more chance. According to Professor Shinya Yamanaka, a notable alumnus and Nobel prize winner, the magic word "VW" (good vision together with hard work) will probably serve as a key to a big success in our life. I would like to believe that this is a very good advice.

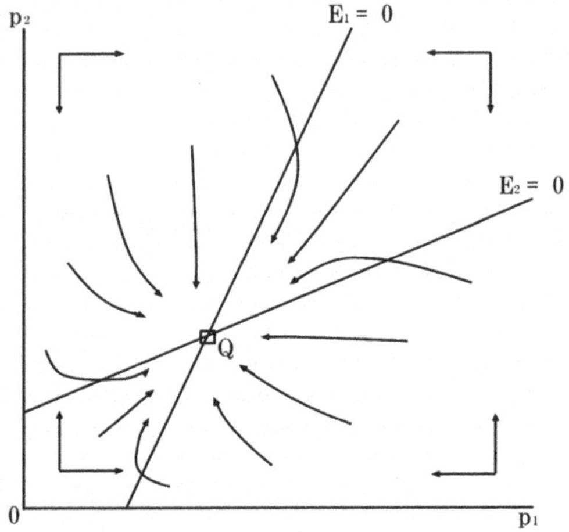

Fig. 8.2 The existence and stability of the two-good markets: An ideal state

8.3 Many Non-normal Cases: Instability, Multiple-Equilibrium, and Nonexistence

As the saying goes, saying is one thing but doing is really another. All the foregoing analyses have just dealt with the ideal, utopian two-good world in which the only one point of general equilibrium uniquely exists and its stability is also guaranteed. One might wonder if such utopian world really exists. When we are getting into the real world as it is, it is far from perfect, with a lot of irregularities. For instance, general equilibrium might not exist at all. There would be more than one equilibrium. Besides, many other "irregular situations" including unstable points and/or saddle points might occur.

8.3.1 The Saddle-Point Equilibrium: Substitutable Goods with Dominating Cross Effects

Let us suppose that there is a market equilibrium with substitutable goods. Then, the most important question to ask is whether and to what extent the stability of the equilibrium is guaranteed. Interestingly enough, the equilibrium is not always stable. In fact, we will see that the cross effects of price changes act as disturbing factors of instability. In fact, if the cross effects are strong enough to outweigh the own effects, then the equilibrium point will become a non-stable saddle point. To clarify such an interesting point, let us focus on the case that the excess demand functions are of the following forms:

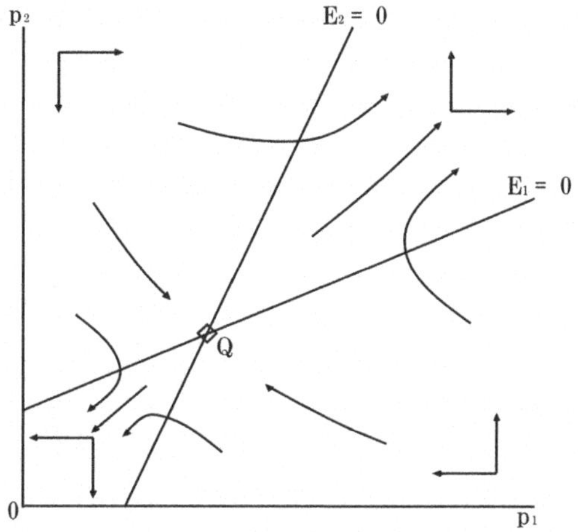

Fig. 8.3 The equilibrium point is an unstable saddle point: The cross effects are stronger than the own effects

$$\boldsymbol{E_1} : E_1 = -p_1 + 2p_2 - 4. \tag{8.9}$$

$$\boldsymbol{E_2} : E_2 = 3p_1 - p_2 - 6. \tag{8.10}$$

The excess demand lines of good 1 and good 2 are depicted in Fig. 8.3. Since we are dealing with substitutable goods, it follows that both lines should be positively sloped. There is an important point we should not miss, however. Note, the cross effects are strong enough to outweigh the own effects. Indeed, in Eq. (8.9) above, the absolute value 2 of the coefficient of p_2 is greater than the one 1 of the coefficient of p_1, whereas in Eq. (8.10), the absolute value 3 of the coefficient of p_1 exceeds the one 1 of the coefficient of $p_{2.}$ As a result, in sharp contrast to the idealistic case discussed above, the slope of E_1 line is steeper for p_2-axis, whereas the one of E_2-line is steeper for p_1-axis.

Let us compare Figs. 8.2 and 8.3. Then, we clearly understand that the positional relationship of the lines $E_1 = 0$ and $E_2 = 0$ has been reversed. Naturally, such reversal of locations would critically change the adjustment processes of prices. Indeed, cross effects are now strong enough to outweigh the own effects. In Eq. (8.9) above, the absolute value 2 of the coefficient of p_2 is greater than the one 1 of the coefficient of p_1, whereas in Eq. (8.10), the absolute value 3 of the coefficient of p_1 exceeds the one 1 of the coefficient of $p_{2.}$ As a result, in sharp contrast to the idealistic case discussed above, the slope of E_2 line is steeper for p_2-axis, whereas the one of E_2 line is steeper for p_1-axis.

Note that the coordinates of equilibrium point are $Q = (16/5, 18/5)$. Apart from the previous discussion, however, we can no longer say that this point is a stable point in its true sense. In fact, it should be a saddle point such that it is unstable on almost all paths. As is seen in Fig. 8.3, except only for the two stable paths—one path going down from the North-West to Q and another going up from the South-East to Q—all other possible paths are not stable at all. This implies that the saddle point Q should substantially be unstable.

8.3.2 Nonexistence of the Two-Market Equilibrium

What we should consider next is the emergence of "non-normal situations." Since all of them are of vital importance, they respectively require very careful investigations. As the first example, let us examine the case in which the excess demand functions are written as follows:

$$\boldsymbol{E_1} : E_1 = -2p_1 + p_2 - 2. \tag{8.11}$$

$$\boldsymbol{E_2} : E_2 = p_1 - 2p_2 - 2. \tag{8.12}$$

Let us take a look at Fig. 8.4. As is seen in Chart (A), the two excess demand lines, $E_1 = 0$ and $E_2 = 0$, never intersect at all in the positive coordinates of the price space. If we dared to solve for p_1 and p_2 in Eqs. (8.11) and (8.12), we would obtain a pair of *negative* prices $(p_1, p_2) = (-2, -2)$, which should not economically be meaningful. Eventually, as is shown in Chart (A), both prices, p_1 and p_2, tend to eventually decline toward zero along the directions of arrows.

If we interchange the positions of the two lines, $E_1 = 0$ and $E_2 = 0$ in the left chart (A), then we immediately obtain those lines in the right chart (B). Then, we have to deal with the following set of equations:

$$\boldsymbol{E_1}\text{:}\ E_1 = -p_1 + 2p_2 + 4. \tag{8.13}$$

$$\boldsymbol{E_2}\text{:}\ E_2 = 3p_1 - p_2 + 6. \tag{8.14}$$

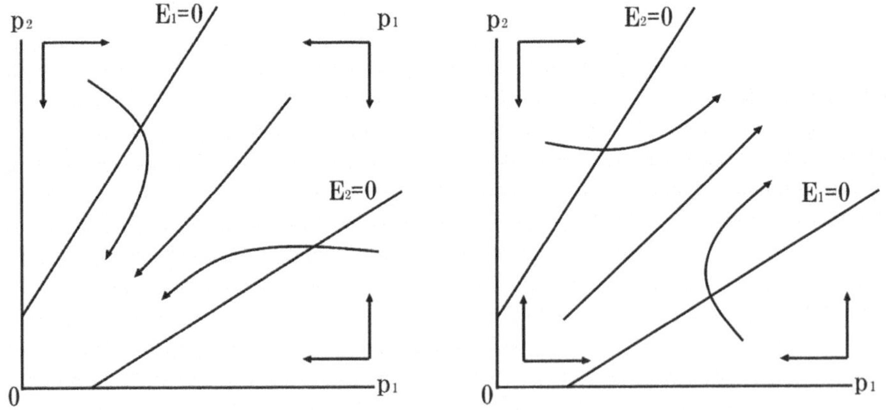

(A) The prices are declining to zero

(B) The prices are rising forever

Fig. 8.4 Nonexistence of equilibriums

In Chart (B), there exist no intersections of the two lines in the positive price quadrant. If we dared to simultaneously solve $E_1 = 0$ and $E_2 = 0$, then we would find $(p_1, p_2) = (-16/5, -18/5)$, a pair of negative and economically meaningless prices. As a result, there are no general equilibrium solutions, and the prices eventually tend to rise forever as indicated by arrows.

8.3.3 The Case of Complementary Goods: A Completely New Situation

In the above, we have limited our investigation to the case of competitive or substitutable goods. Looking at the real world, however, we can eyewitness many other cases in which some goods may not be mutually competitive but rather move in the same direction. For instance, while tea and coffee are substitutes, tea and sugar are complementary, and so are coffee and sugar.

As the first example of complementary goods, let us discuss the following excess demand functions of the two goods:

$$\boldsymbol{E_1} : E_1 = -2p_1 - p_2 + 4. \tag{8.15}$$

$$\boldsymbol{E_2} : E_2 = -p_1 - 3p_2 + 6. \tag{8.16}$$

Let take a close look at Eqs. (8.15) and (8.16). Then, we will see that in the right-hand sides of those equations, the values of coefficients of p_1-term and p_2-term are both negative. This means that both the excess demands for x_1 and x_2 have to increase when p_1 falls or when p_2 falls. This implies that the two goods are complementary, not substitutes. Besides, we assume here that the own effects overpower the cross effects. Therefore, we would rightly conjecture that the equilibrium point Q, if it exists, must be stable. Surely, such conjecture will turn out be right by looking at the left chart (A) of Fig. 8.5.

Concerning Fig. 8.5, there is one more point to watch. We must make sure that the two excess demand lines, E_1 and E_2, are now *negatively* sloped. Since the own effects dominate the cross effects in Chart (A), it follows that both the slope of E_1 for p_1-axis and the slope of E_2 for p_2-axis should be steeper.

We will turn to the second case, which is represented by the following pair of excess demand functions:

$$\boldsymbol{E_1} : E_1 = -p_1 - 3p_2 + 6. \tag{8.17}$$

$$\boldsymbol{E_2} : E_2 = -2p_1 - p_2 + 4. \tag{8.18}$$

Now, the cross effects are now more powerful than the own effects. As can easily be seen in Chart (B), the slope of E_1 for p_2-axis (not for p_1-axis) is steeper while the

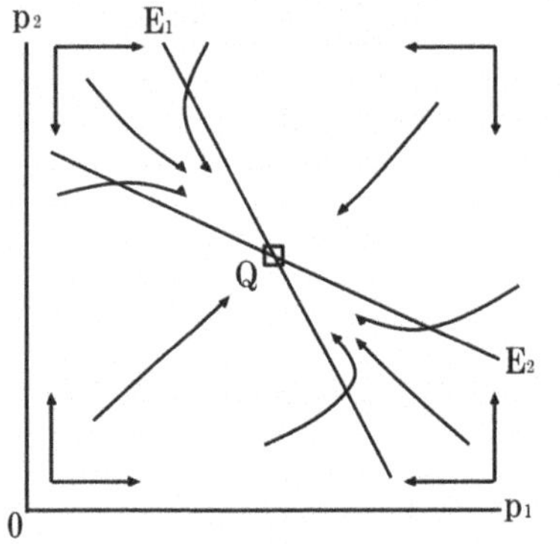

(A) Point Q is stable since the own effects are dominating

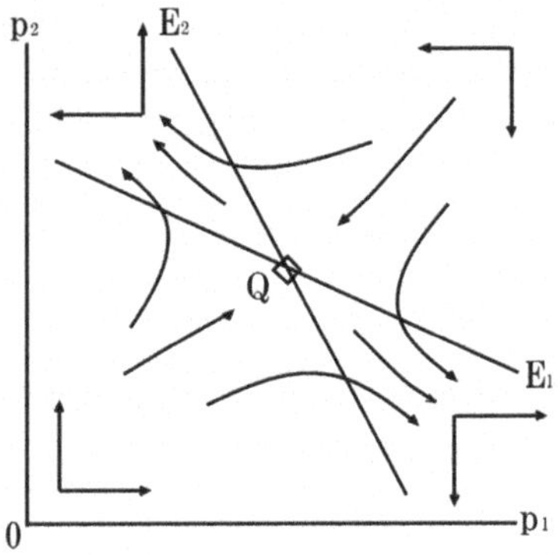

(B) Point Q is unstable since the cross effects are dominating

Fig. 8.5 The two goods are complementary

one of E_2 for p_1-axis (not for p_2-axis) is steeper. Therefore, the equilibrium point Q is an almost unstable saddle point.

In conclusion, the power relation between the own effects and the cross effects plays a vital role in the determination of the stability of an equilibrium point. Whether the goods are substitutes or complementary, the dominance of the own effects over the cross effects clearly guarantees the stability of an equilibrium. On the contrary, if the cross effects dominate the own effects, then an equilibrium point becomes an almost unstable saddle point.

8.4 Comparative Statics: The Impact of Parameter Changes on Equilibriums

Let us turn our attention to the issue of comparative static analysis. This is a very important problem that was first intensively discussed by J.R. Hicks (1939, 2nd. ed. 1946) and later greatly developed by Michio Morishima (1984).

First of all, let us assume that the two goods are substitutes like tea and coffee. Let us take a look at Fig. 8.6. In the left chart (A), the straight lines E_1 and E_2 respectively represent the excess demand lines of good 1 (tea) and good 2 (coffee). More specifically, let us consider the following set of equations:

$$\boldsymbol{E_1} : E_1 = -2p_1 + p_2 + 4. \tag{8.19}$$

$$\boldsymbol{E_2} : E_2 = p_1 - 3p_2 + 6. \tag{8.20}$$

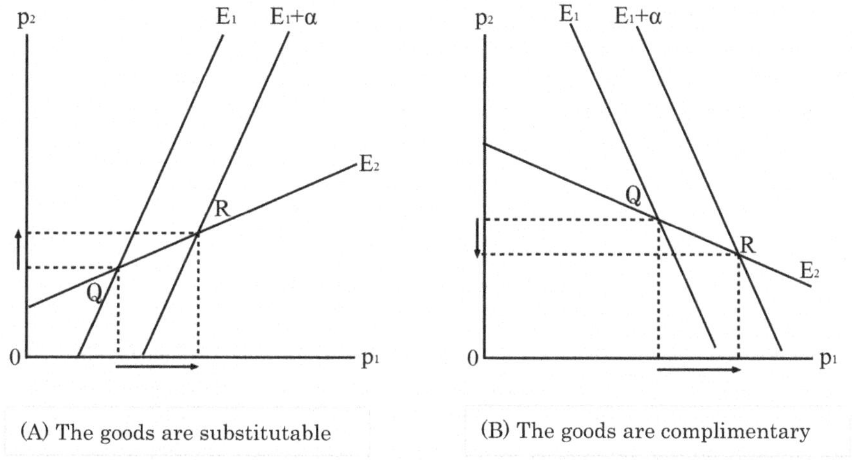

Fig. 8.6 Comparative statics: changes in excess demand

If we put $E_1 = 0$ and $E_2 = 0$ and solve for p_1 and p_2, we find a pair of equilibrium prices: in fact, $(p_1{}^{Q}, p_2{}^{Q}) = (3.6, 3.2)$.

Now suppose that because of some reasons, the popularity of tea suddenly rises among the people. Then, the excess demand curve E_1 has to shift to the right. This implies that regarding the constant term of Eq. (8.1) above, it has to change its value from 4 to (4 + α):

$$\textbf{A change in } \boldsymbol{E_1} : E_1 + \alpha = -2p_1 + p_2 + 4 + \alpha. \tag{8.21}$$

Let us take a look at Fig. 8.6. In left chart (A), the new equilibrium point R is found at the intersection of the two lines, $E_1 = 0$ and $E_2 = 0$. For convenience, let us specify $\alpha = 2$. Then, the coordinates of R are calculated as $(p_1{}^{R}, p_2{}^{R}) = (4.8, 3.6)$.

The case in which the two goods are no longer substitutable but rather complementary is well represented by the right chart (B) in Fig. 8.6. A comparative static analysis for such a case will be very interesting and effectively be carried out a bit later.

8.4.1 Hicksian Three Laws of Price Changes: The Case of Substitutable Goods

In his classic *Value and Capital,* Hicks (1939, 2nd ed. 1946) succeeded in deriving the very famous three laws, which stood for the core of his comparative static results. It was also remarkable to see that those Hicksian laws were later developed a great deal by his successor, Morishima (1984). It is the Hicksian laws that we will intensively investigate in this and following subsections.

We are most interested in comparing the two equilibrium points, the old point Q and the new point R. We are expected to obtain several results of great importance. First of all, we would like to pay attention to comparison of the following two relations:

$$p_1{}^{R} = 4.8 > 3.6 = p_1{}^{Q}. \tag{8.22}$$

$$p_2{}^{R} = 3.6 > 3.2 = p_2{}^{Q}. \tag{8.23}$$

To begin with, let us suppose that because of some reasons, tea becomes very popular among the people, whence its demand suddenly increases. To take an example, the opening of the Great Britain Exhibition in Kyoto may produce such a sudden surge for the popularity of tea in Japan. Then, its immediate effect we can expect to have is that tea will be more expensive than before in response to the demand increase. Not only that, there would also be many other repercussions such as its indirect impact on other markets including the coffee market. Besides, those influences will invoke the third and fourth rounds of repercussions. Those direct and indirect effects would make our repercussion calculations very hard and complicated. In order to make sure of this point, if we patiently compute the *proportional rates of increase* of the prices (in terms of percent), then we will be able to derive the following equations:

$$\left(p_1{}^{R} - p_1{}^{Q}\right)/p_1{}^{Q} = (4.8 - 3.6)/3.6 = 1/3. \tag{8.24}$$

$$\left(p_2{}^{R} - p_2{}^{Q}\right)/p_2{}^{Q} = (3.6 - 3.2)/3.2 = 1/8. \tag{8.25}$$

Clearly, the ratio $(p_1{}^{R} - p_1{}^{Q})/p_1{}^{Q}$ is greater than the ratio $(p_2{}^{R} - p_2{}^{Q})/p_2{}^{Q}$, implying that the rising rate of tea exceeds the one of coffee.

We must bear in mind that whether or not the excess demand functions are linear, all the comparative static results derived above generally hold. J.R. Hick first obtained all those results that Michio Morishima rightfully called the "Hicksian Three Laws of Price Changes." They are really outstanding consequences of the Hicks–Morishima approach, which can be summarized as the following theorem.

Theorem 8.1 (Hicks–Morishima on Substitutable Goods) Assume that the two goods, x_1 and x_2, are substitutable and that the own effects overpower the cross effects. Then the point of general equilibrium, if it exists, must be stable. When the excess demand for good 1, E_1, increases, the following three properties must hold:

1. The price of good 1, p_1, will rise.
2. The price of good 2, p_2, will rise too.
3. The price of good 1 will rise proportionately more than the price of good 2; namely, the price ratio (p_1/p_2) will rise.

8.4.2 The Case of Complementary Goods: Hicksian Three Laws Reformulated

Now, let us turn to the case of complementary goods. Let us assume that goods 1 and 2 respectively represent tea and sugar. Then, as the right chart (B) of Fig. 8.6 above tells us, the two excess demand lines, E_1 and E_2, are both downward sloped. If we suppose that the own effects outweigh the cross effects, then the old intersection point Q becomes a stable point.

Now, suppose that a sudden tea boom happens among the people, resulting in a drastic increase in tea demand. This is the first direct effect of tea price rise. Moreover, such rise will lead to a sequence of secondary repercussions since it will dampen the demands for both tea and sugar. Recall here that complementary goods have to go in the same direction. This is the second indirect effect. If we employ a similar reasoning, then we would also take account of third and fourth round of repercussions that must have less influence than the second round.

In Chart (B), let us shift the excess demand function to the right, from E_1 to $E_1 + \alpha$. Then, since such shift changes the equilibrium point in a right and downward direction, from Q to R, it will have to cause a big rise in p_1 together with a small change in p_2. Consequently, in absolute value terms, the rising rate of p_1 will exceed the falling rate of p_2. We omit a more detailed calculation here.

The final results for the case of complementary goods will be summarized as the following theorem:

Theorem 8.2 (Hicks–Morishima on Complementary Goods) Suppose that the two goods, x_1 and x_2, are complementary, and that the own effects overpower the cross effects. Then, the point of general equilibrium, if it exists, must be stable. When the excess demand for good 1, E_1, increases, the following three properties must hold:

1. The price of good 1, p_1, will rise.
2. The price of good 2, p_2, will fall a bit.
3. The rising rate of good 1 will be greater than the falling rate of good 2; namely, price ratio (p_1/p_2) will rise a great deal.

As I mentioned before, the exact derivation of Theorems 8.1 and 8.2 was the main theme of the Young Morishima. For that purpose, he literally racked his brains. In this chapter, I have intended to limit my analysis on a simple yet useful case of *linear* excess demand functions. Besides, following the Morishima spirits, I have often made use of many eye-catching graphical approaches. I can say, however, that the extension of my analysis to the more general nonlinear cases should not be so difficult, requiring a bit more calculation and patience.

8.5 Morishima Toward a New Economics

This chapter may be regarded as a homage to my two great mentors, Professors John W. Hicks and Michio Morishima. The Hicks–Morishima approach was adopted and developed by Morishima in his book (Morishima 1984). For this book, Morishima gave an interesting subtitle "An Introduction to New Economics." Because it was far more than "an introduction" and contained "several new and original results," it seemed to me that the subtitle per se was "too modest, and even cheating." In fact, Morishima himself made several remarks:

> The approach of this book differs to a considerable degree from standard economics. . . . This book attempts to analyze the price mechanism in accordance with reality, and at the same time to introduce students directly to the major problems of economics an analysis of the way in which the real economy operates and the best way to bring about a change in direction in this operation. (Morishima 1984, Preface)

The new books of Morishima (1999a, b) could be thought of as one of his "last swans," in which he greatly lamented the recent Japanese society as one approaching at a "dangerous stage." Here again, he made some intriguing remarks:

> This book is the one I have constantly wanted to write for a long time. In fact, it is a sort of a "Grand Integral Social Science" in the sense that it contained economics, sociology, education, history and many other related fields. It could rightly be called an "Academic Symphony." ...... Let us take a careful look at many non-economic fields such as the worlds of politics, religion, and ideology. Then you will see that many main actors such as political, religious, and military bosses may play very large roles in our society. Unfortunately, economics and other related fields are so powerless to investigate those important problems. So, I should say that the existing social science remains in an under-developed state, thus lacking its very essential core. (Morishima 1999b, Appendix)

Honestly speaking, the right way toward the "Grand Integral Social Science" a la Morishima seems to be too far for me to take. Even a right direction for the road has not been found yet. A few years ago, when I published my book, namely Sakai (2010) *The Economic Thought of Risk and Uncertainty* written in Japanese, I sent one copy to Professor Tetsuya Nosse, a close friend of Morishima. Nosse immediately replied to me, "This looked like 'an academic symphony' a la Morishima!" This was really the highest praise on my academic work. Quite recently, I published a new English version of *Economic thought of risk and uncertainty,* namely Sakai (2019) *J.M. Keynes versus F.H. Knight: risk, probability, and uncertainty.* It is my sincere wish that it has been well received by Morishima and Nosse.

I heartily hope that this chapter would serve very well as one small step toward Morishima's dream of establishing the Grand Integral Social Science. I believe in the following golden saying: "Life is an adventure! Life is really a challenge!"

References

Arrow KJ, Debreu G (1954) Existence of equilibrium for a competitive economy. Econometrica 22: 265–290

Arrow KJ, Hahn FH (1971) General competitive analysis. Holden-Day, San Francisco

Debreu G (1959) Theory of value. Wiley, New York

Hicks J (1939, 2nd ed. 1946) Value and capital. Oxford University Press, Oxford

Hirota M, Sakai Y (1969) Notes on substitution and scale effects in production theory, unpublished manuscript. University of Rochester

Komoro N (2004) Economics super stars: seminar on the history of economic thought (in Japanese). Diamond Publishers, Tokyo

McKenzie LW (1954) On equilibrium in Graham's model of world trade and other competitive systems. Econometrica 22:239–257

McKenzie LW (1959) On the existence of general equilibrium for a competitive market. Econometrica 27:54–71

Morishima M (1974) Marx in the light of modern economic theory. Econometrica 42(4):611–632

Morishima M (1984) The economics of industrial society. Cambridge University Press, Cambridge

Morishima M (1999a) The life and creed of an economist (in Japanese). Iwanami Publishers, Tokyo

Morishima M (1999b) Why will Japan collapse? (in Japanese). Iwanami Publishers, Tokyo

Sakai Y (1972) Axiomatic foundations of consumption and production theories. Ph.D. dissertation thesis, University of Rochester

Sakai Y (2010) The economic thought of risk and uncertainty (in Japanese). Minerva Publishers, Kyoto

Sakai Y (2019) J.M. Keynes versus F.H. Knight: risk, probability, and uncertainty. Springer

Chapter 9
International Trade Under Risk: Comparative Static Analysis

Abstract The risk-free, two-sector, two-factor, constant-returns-to-scale model has long served as a standard model of international trade, with R. W. Jones and M.C. Kemp among others being its key promoters. The purpose of this chapter is to make an effort to extend the basic theorems of the standard model to cover new situations with price risk. The question to ask is whether and to what extent those results are still applicable to the world under risk. It is shown that when firms in the risky factor exhibit decreasing absolute risk aversion, the Rybczynski theorem and the Stolper–Samuelson theorem may fail to hold for some cases, whereas the factor price equalization theorem cannot carry over to the stochastic world. Besides, the implications of uniform (relative) changes in both factor endowments, and those in both (expected) commodity prices, are carefully investigated.

Keywords Standard model of international trade · Decreasing absolute risk aversion · Comparative static results · Rybczynski theorem · Stolper–Samuelson theorem · Factor price equalization theorem

9.1 The Basic International Trade Model and Its Possible Extension to the World of Risk and Uncertainty

In the past, a simple general equilibrium model of production has long served as an important workhorse for the evolution of many areas of economics, including international economics, public finance, and economic growth. Presumably, the most famous model of this sort is the risk-free, two-sector, two-factor, constant-returns-to-scale model. In this chapter, we attempt to blend the more recent development in the economics of risk and uncertainty with some important problems

---

This chapter is a completely revised version, based on Sakai (1978) and Sakai (1982). I am grateful to Ronald Jones (Rochester), Michihiro Ohyama (Keio), Murray Kemp (Australian National), Jose Sheinkman (Chicago), Katsuhiko Suzuki (Kwansei Gakuin), Akihiro Amano (Kobe), and Akira Takayama (Purdue) for many valuable suggestions.

Y. Sakai, *Games, Decisions, and Markets*, New Frontiers in Regional Science: Asian Perspectives 74, https://doi.org/10.1007/978-981-99-5285-4_9

which has been traditionally examined within the framework of the risk-free, two-by-two model.[1]

We are concerned with the competitive economy in which one of the two sectors is confronted with risk as the result of price fluctuations in its product. Extending the usual, small-country assumption to the price risk milieu, we assume that firms in the risk-affected sector have no control over the distribution function of the price of their product, whose shape depends on world weather, people's taste patterns, and many other natural and socioeconomic factors influencing the demand–supply conditions in the world-wide market. Following Arrow and Pratt. we further assume that those risk-affected firms exhibit decreasing absolute risk aversion and seek to maximize expected utility from profits.[2]

In deriving comparative static results from the aforementioned model under price risk, it is very effective for us to highlight the dual relationship between factor endowments and commodity outputs on the one hand and factor prices and commodity price parameters on the other. Indeed, it may prove so convenient for this purpose to make full use of the variable, input–output coefficients of both sectors instead of the production functions per se. Clearly, seeing to what extent the presence of price risk affects the dual structure of the model and thus forces the well-known, non-stochastic results to be amended should be the main task assigned to the present study.[3]

More specifically, we will examine the possibility of extending the basic theorems of the traditional, risk-free model to cover situations with price risk. It is recalled here that, in the risk-free world, the profits of each firm must vanish at the long-run equilibrium. As will be seen below, however, even at such a long-run equilibrium, the risk-affected sector's expected profits should be positive in the case of risk aversion. Accordingly, the part of the expected price of the risk-affected output exceeding the wage rate and rental may be regarded as the per unit payment for the third hypothetical "risk factor." Existence of such "risk-bearing fee" and its responsiveness to changes in factor endowments and in commodity price parameters should require special care in exploring the comparative static properties of our stochastic model.

---

[1]For the standard treatment of the non-stochastic, two-by-two model, see Caves and Jones (1973). Jones (1965), Bhagwati (1964), Komiya and Amano (1972), Takayama (1972), Feenstra (2004), Ohyama (2011), Kemp (1969, 1976), Leamer (1995), and Krugman and Obstfeld (2007).

[2]Empirical analysis has shown that many countries suffer from high price fluctuations in their exporting commodities. See Coppock (1962), Komiya and Amano (1972), Krugman and Obstfeld (2007), and Ohyama (2011).

[3]The basic characteristics of the two-by-two equilibrium model under uncertainty was investigated by Batra (1974, 1975a, 1975b). His interest was mainly centered around *production* uncertainty rather than *price* uncertainty. Unfortunately, his formation of the model was fairly tedious and obscured the duality feature of the model. Although Mayer (1976) developed an interesting model similar to our model here, he was merely content to work with the *absolute* rate of change in a variable such as dK. Following the Rochester tradition led by Jones (1965), Caves and Jones (1973), Komiya and Amano (1972), Takayama (1972), and Ohyama (2011), a more powerful mathematical tool of the *relative* rate of change such as $d \log K$ $(=dK/K)$ was efficiently employed in our paper.

The following results among other things are worth mentioning here.

1. Although the powerful factor-price equalization theorem can be established in the risk-free world, it is no longer valid—at least in its strict version—in the stochastic context.
2. The famous Rybcznski and Stolper–Samuelson theorems may fail to hold for certain cases to be specified below. So, this partially invalidates the Heckscher–Ohlin theorem as well.
3. A uniform relative change in the two factor endowments affects commodity outputs unevenly, whereas a uniform relative in the two (expected) commodity prices affects factor prices unevenly.
4. The nice magnification effect advocated by Ronald Jones cannot carry over to the stochastic world.

The contents of this chapter are as follows. In Sect. 9.2, we will make a set of assumptions to establish our two-sector two-factor model with price risk. In Sect. 9.3, we will place the key equations describing the model on a common footing by rewriting them in terms of relative changes in variables and parameters. Section 9.4 will discuss the relationship between factor prices and commodity price parameters. The subsequent sections will be devoted to the comparative static properties of the model. Namely, we will explore the implications of changes in factor endowments in Sect. 9.5, those of changes in (expected) commodity prices in Sect. 9.6, and those of changes in price risk in Sect. 9.7. And final remarks will be made in Sect. 9.8.

9.2 Basic Assumptions and the Model with Price Risk

Throughout this chapter, we maintain the following set of basic assumptions under which our stochastic trade model is to be established. As is easily recognized, some of those assumptions are usually made for the traditional risk-free model, whereas several others are newly installed to take account of price risk.

More specifically, the following set of assumptions and notations will be maintained.

1. Two primary factors, labor (L) and capital (C), are used in producing two distinct commodities, the uncertainty or risk-affected commodity (U) and the certainty or risk-free commodity (C). Technology in each of the two sectors exhibits constant returns to scale.
2. The money wage rate and the money rent for capital service are respectively denoted by w and r, whereas the market prices of the two commodities are denoted by p_U and p_C.
3. Firms in sector U face price risk, but firms in sector C are risk-free. Free competition prevails in all markets. The provability distribution of p_U, together with p_C, is exogenously given to the country in question; in other words, the "small country" assumption in the probabilistic sense is still maintained.

4. Factors are incessantly supplied and fully employed. While factors are perfectly mobile between the two sectors, factor intensity rankings never reverse even when factor prices change.
5. On the one hand, firms in sector U, which have to make their input–output decision prior to the knowledge of the price of their output, seek to maximize expected utility from profits. It is assumed that those firms are not only risk averters but also exhibit decreasing absolute risk aversion. On the other hand, firms in sector C seek to maximize utility from profits.
6. All quantities should be valued in terms of money. Money is implicitly present in the economy and serves merely as the unit of measurement.
7. The demand side and other socio-politico-economic sides that must be introduced for a fuller description of the economy are all but neglected here.

With those assumptions in mind, let us build a two-sector, two-factor general equilibrium model with price risk. First of all, we note that the linearly homogeneous production function and the profit level in sector U are respectively given as follows.

$$U = F_U\ (L_U, K_U).$$

$$\Pi_U = p_U\ U - w\ L_U - r\ K_U.$$

The random variable p_U can be rewritten as follows.

$$p_U = \mu_U + \gamma\varepsilon,$$

where $E\ [\varepsilon] = 0$, and γ is a shift parameter, which is assumed to be 1 initially. An increase in γ leads to an increased spread of the probability distribution of p_U around the constant mean μ_U. Clearly, this may be regarded as the definition of an increase in the risk of p_U.[4]

Let V_U be the utility attainable from Π_U. Assuming risk aversion on the part of U producers, we have the following equations:

$$V_U{}'\ (\Pi_U) > 0,\ V_U{}''(\Pi_U) < 0.$$

Those producers are interested in maximizing $E\,[\![V_U\ (\Pi_U)]\!]$ with respect to their decision variables, L_U and K_U. This results in the following equations:

$$E[\![V_U{}'\ (\Pi_U)\ \{(\mu_U + \gamma\varepsilon)\ F_{LU} - w)\}]\!] = 0, \tag{9.1}$$

[4]It is assumed here that p_U is positive for any value of ε and that $\Phi(\varepsilon)$ takes on positive values for any ε. Therefore, both $E\ [p_U]$ and $Var\ [p_U]$ should be positive. For such mean-preserving spread type of shift in the distribution function, see Sandmo (1971) and Sakai (1977).

$$E[V_U{}'(\Pi_U)\{(\mu_U+\gamma\varepsilon)F_{KU}-r\}]=0, \tag{9.2}$$

where we note[5]

$$F_{LU}\equiv \partial F_U/\partial L_U, F_{KU}\equiv \partial F_U/\partial K_U.$$

From (9.1) and (9.2), we immediately obtain the following equations:

$$w=(\mu_U-\gamma\rho)F_{LU}. \tag{9.3}$$

$$r=(\mu_U-\gamma\rho)F_{KU}. \tag{9.4}$$

Note that ρ is defined as follows.

$$\rho\equiv -E[V_U{}'(\Pi_U)\,\varepsilon]/E[V_U{}'(\Pi_U)]. \tag{9.5}$$

The value of ρ is really important and requires careful interpretation. It represents the per unit psychological cost of risk-bearing associated with price risk, which may reasonably be called the "risk-bearing fee." As will be shown in Lemma 9.1 below, ρ is always positive whenever U producers are risk-averse. Therefore, a nice interpretation may be given to Eqs. (9.3) and (9.4): At U producer's risk-affected equilibrium, the price of each factor is equal to the "net" or "effective" expected price of its marginal product. Since factor prices are positive, μ_U must be greater than $\gamma\rho$.

Lemma 9.1 (Risk-Bearing Fee) **If $V_U{}''(\Pi_U)<0$, then $\rho>0$.**

Proof From the definition of Π_U, we immediately find $\Pi_U - E[\Pi_U] = \gamma\varepsilon U$. Therefore, if $V_U{}''(\Pi_U)<0$, we have the following relationship:

$$V_U{}'(\Pi_U)\lesseqgtr V_U{}'(E[\Pi_U])\Leftrightarrow \gamma\gtreqless 0.$$

This immediately implies the following:

$$V_U{}'(\Pi_U)\,\varepsilon\; V_U{}'(E[\Pi_U])\,\varepsilon \text{ for any value of } \varepsilon, \tag{9.6}$$

with the equality only when $\varepsilon = 0$. Hence, taking expectations on both sides of Eq. (9.6), we obtain

[5] The second-order condition for a regular maximum requires that the Hessian matrix of $E[V_U(\Pi_U)]$ with respect to L_U and K_U be negative definite. This condition is weaker than the condition that the Hessian matrix of $F_U(L_U, K_U)$ be negative definite, provided U firms are risk-averse. For this point, see Sakai (1977), p. 33.

$$E[V_U'(\Pi_U)\,\varepsilon] < V_U'(E[\Pi_U])\,E[\varepsilon] = 0.$$

In the light of Eq. (9.5), this implies that ρ must be positive.
Q.E.D.

Concerning the behavior of C producers, they are interested in maximizing utility from profit:

$$V_C(\Pi_C) = V_C(p_C F_C(L_C, K_C) - w L_C - r K_C),$$

where V_C is the increasing utility function of Π_C, and F_C the linearly homogeneous production function of (L_C, K_C). Then, the optimum conditions are given as follows:

$$w = p_C F_{LC}, \tag{9.7}$$

$$r = p_U F_{KC}, \tag{9.8}$$

where $F_{LC} \equiv \partial F_C / \partial L_C$, $F_{KC} \equiv \partial F_U / \partial K_C$. In the light of Eqs. (9.3), (9.4), (9.7), and (9.8), we have the following equation.

$$\frac{F_{LU}}{F_{KU}} = \frac{w}{r} = \frac{F_{LC}}{F_{KC}}. \tag{9.9}$$

Therefore, as with the certainty case, the factor price ratio must be equal to the marginal product ratio at each sector.

We note that the technology of the economy can also be described by the following matrix:

$$a = \begin{bmatrix} a_{LU} & a_{LC} \\ a_{KU} & a_{KC} \end{bmatrix}.$$

where a_{ij} stands for the quantity of factor i required to produce a unit of commodity j; namely, $a_{LU} \equiv L_U / U$, $a_{KU} \equiv K_U / U$, etc.

Let us try to reformulate our two-by-two model in terms of the technology matrix (a_{ij}) instead of the production functions F_U and F_C. First of all, the requirement that both factors be fully employed is obviously given as follows.

$$L = a_{LU} U + a_{LC} C. \tag{9.10}$$

$$K = a_{KU} U + a_{KC} C. \tag{9.11}$$

Secondly, noting that the production function F_C of the uncertainty sector is linearly homogeneous, application of the Euler theorem yields the following equation.

$$U = F_{LU}L_U + F_{KU}K_U,$$

which leads to the following equation.

$$\mu_U - \gamma\rho = (\mu_U - \gamma\rho)\, F_{LU}\, a_{LU} + (\mu_U - \gamma\rho)\, F_{KU}\, a_{KU}.$$

In view of Eqs. (9.3) and (9.4), this results in the following.

$$\mu_U = wa_{LU} + ra_{KU} + \gamma\rho. \tag{9.12}$$

In a similar way, we can obtain the following equation for the certainty sector.

$$p_C = wa_{LC} + ra_{KC}. \tag{9.13}$$

As is quite clear from comparison between Eqs. (9.10) and (9.11) and Eqs. (9.12) and (9.13), the present formulation based on the input–output matrix (*aij*) is very instructive in showing the formal similarity of the relationship between factor endowments and commodity outputs on the one hand and relationship between commodity price parameters and factor prices on the other. Such a duality feature will be proved much more deeply in carrying out comparative static analysis in the subsequent sections.

In the general case of variable coefficient, we note that each *aij* depends on the rate of factor prices in the following way:[6]

$$\begin{aligned} a_{LU} &= a_{LU}(w/r), a_{KU} = a_{KU}(w/r), \\ a_{LC} &= a_{LC}(w/r), a_{KC} = a_{KC}(w/r). \end{aligned} \tag{9.14}$$

It is now a simple matter to show that the expected profits of U producers are positive if those producers are risk-averse. Indeed, in view of Eq. (9.12), we find the following equation.

$$\begin{aligned} \Pi_U &= (\mu_U + \gamma\varepsilon)U - w\, a_{LU}\, U - r\, a_{KU}\, U \\ &= \{(\mu_U + \gamma\varepsilon) - (w\, a_{LU} + r\, a_{KU})\}U \\ &= \{(\mu_U + \gamma\varepsilon) - (\mu_U - \gamma\rho)\}U \\ &= (\rho + \varepsilon)\gamma U. \end{aligned} \tag{9.15}$$

[6]Linear homogeneity of the production function F_U $(L_U, K_U) = U$ implies F_U $(a_{LU}, a_{KU}) = 1$. Besides, since the first derivatives F_{LU} and F_{KU} are homogeneous of degree zero, it follows from Eq. 9.9 that F_{LU} $(a_{LU}, a_{KU})/F_{LU}$ $(a_{LU}, a_{KU}) = w/r$. Hence, provided F_U is well-behaved, these two equations yield $a_{LU} = a_{LU}$ (w/r) and $a_{KU} = a_{KU}$ (w/r). Similarly, we can obtain $a_{LC} = a_{LC}$ (w/r) and $a_{KC} = a_{KC}$ (w/r).

We thereby have $E\,[\Pi_U] = \gamma\rho U$, implying that expected profits are positive whenever $V_U''\,(\Pi_U)$ is negative. This is in marked contrast to the certainty sector in which profits are of course zero.[7]

The production structure in the economy is thus determined by ten independent equations including (9.5) and (9.10)–(9.15). (Note that the system (9.14) per se contains four equations. Our uncertainty model is expected to determine ten variables: U, C, w, r, a_{LU}, a_{KU}, a_{LC}, a_{KC}, ρ, and Π_U. Hence, we have a determinate system with L, K, μ_U, γ, and p_C being treated as parameters.

9.3 The Equations of Change

We are now in a position to explore the comparative static properties of the model introduced in the previous section. In so doing, it will prove quite useful to place the key equations describing our model on a common footing by rewriting them in terms of *relative or proportional changes* in variables and parameters.

In what follows, we will effectively make use of *log differentiation* rather than simple differentiation. For instance, we note that $d\log L = dL/L$, $d\log w = dw/w$, $d\log a_{KU} = da_{KU}/a_{KU}$, etc. If we differentiate (9.10) and rearrange it, then we find the following equation.

$$a_{LU}\,dU + a_{LC}\,dC = dL - (da_{LU}U + da_{LC}C).$$

Dividing both sides of this equation by L and rearranging it results in the following equation.

$$\begin{aligned}&(a_{LU}\,U/L)(dU/U) + (a_{LC}\,C/L)(dC/C)\\ &= dL/L - \left[(da_{LU}/a_{LU})(a_{LU}\,U/L) + (da_{LC}/a_{LC})(a_{LC}C/L)\right].\end{aligned}$$

This equation can be rewritten in terms of logarithmic differential as follows.

$$\begin{aligned}&\lambda_{LU}\,d\,\log\,U + \lambda_{LC}\,d\,\log\,U\\ &= d\,\log\,L - (\lambda_{LU}\,d\,\log\,a_{LU} + \lambda_{LC}\,d\,\log\,a_{LC}),\end{aligned} \tag{9.16}$$

where $\lambda_{LU} \equiv a_{LU}\,U/L$ and $\lambda_{LC} \equiv a_{LC}\,C/L$. Clearly, the λ's refer to factor endowment's *relative* allocations in each sector. A fraction of labor force L is allocated to the uncertainty sector (λ_{LU}), and this fraction plus the fraction of labor force allocated to the certainty sector (λ_{LC}) must add to unity:

[7] For this point, see Sandmo (1971).

$$\lambda_{LU} + \lambda_{LC} = 1. \tag{9.17}$$

In a similar fashion, if we differentiate (9.11) and rearrange it, we can derive the following equation.

$$\begin{aligned} &\lambda_{KU}\, d\, \log\, U + \lambda_{KC}\, d\, \log U \\ &= d\, \log\, K - (\,\lambda_{KU}\, d\, \log\, a_{KU} + \lambda_{KC}\, d\, \log\, a_{KC}), \end{aligned} \tag{9.18}$$

where $\lambda_{KU} \equiv a_{KU}\, U/K$ and $\lambda_{KC} \equiv a_{KC}\, K/L$. We also note that the following equation must hold.

$$\lambda_{KU} + \lambda_{KC} = 1. \tag{9.19}$$

Similarly, the equilibrium price-cost equations (9.12) and (9.13) can also be rewritten as the following equations of *relative* rates of exchange.

$$\begin{aligned} &\theta_{KU}\, d\, \log\, w + \theta_{KU}\, d\, \log\, r + \theta_R\, d\, \log\, \rho \\ &= d\, \log\, \mu_U - \theta_R\, d\, \log\, \gamma - (\theta_{LU}\, d\, \log\, a_{LU} + \theta_{KU}\, d\, \log\, a_{KU}), \end{aligned} \tag{9.20}$$

$$\begin{aligned} &\theta_{LC}\, d\, \log\, w + \theta_{KC}\, d\, \log\, r \\ &= d\, \log\, p_C - (\theta_{LC}\, d\, \log\, a_{LC} + \theta_{KC}\, d\, \log\, a_{KC}), \end{aligned} \tag{9.21}$$

where $\theta_{LU} \equiv a_{LU}\, w/\mu_U$, $\theta_{KU} \equiv a_{KU}\, r/\mu_U$, $\theta_R \equiv \gamma\rho/\mu_U$, $\theta_{LC} \equiv a_{LC}\, w/p_C$, and $\theta_{KC} \equiv a_{KC}\, r/p_C$. The θ's represent factor's *relative* shares in each sector. Hence, θ_{LU} and θ_{KU} respectively stand for labor's share and capital's share in the uncertainty sector, whereas θ_R shows what we may call "risk factor's share" in the same sector. Obviously, these three shares must add to unity:

$$\theta_{LU} + \theta_{KU} + \theta_R = 1. \tag{9.22}$$

In a similar way, we can also find the following equation.

$$\theta_{LC} + \theta_{KC} = 1, \tag{9.23}$$

which shows the zero profit condition for the certainty sector, with no risk factor being present. As far as all physical factors are employed and firms are risk-averse, each θ should be positive and less than unity.

Let us recall that the technology matrix in the economy is given as follows.

$$a = \begin{bmatrix} a_{LU} & a_{LC} \\ a_{KU} & a_{KC} \end{bmatrix}.$$

Since the determinant of the technology matrix is written as $A = |a|$, we find $A = a_{LU}\, a_{KC} - a_{LC}\, a_{KU}$. Obviously, the value of A is positive if the uncertainty sector is relatively more labor-intensive (i.e., $a_{LU}/a_{KU} > a_{LC}/a_{KC}$), it is negative if it is relatively more capital-intensive (i.e., $a_{LU}/a_{KU} < a_{LC}/a_{KC}$). Further, let us define the allocation matrix λ and the share matrix θ as follows:

$$\lambda = \begin{bmatrix} \lambda_{LU} & \lambda_{LC} \\ \lambda_{KU} & \lambda_{KC} \end{bmatrix}, \quad \theta = \begin{bmatrix} \theta_{LU} & \theta_{LC} \\ \theta_{KU} & \theta_{KC} \end{bmatrix}.$$

In the light of (9.17), (9.19), (9.22), and (9.23), we find the following equations.

$$\begin{aligned} \Lambda \equiv |\lambda| &= \lambda_{LU}\lambda_{KC} - \lambda_{LC}\lambda_{KU} \\ &= \lambda_{LU} - \lambda_{KU} = \lambda_{KC} - \lambda_{LC}. \end{aligned} \tag{9.24}$$

$$\begin{aligned} \Theta \equiv |\theta| &= \theta_{LU}\theta_{KC} - \theta_{LC}\theta_{KU} \\ &= \theta_{LU} - (1-\theta_R)\theta_{LC} = (1-\theta_R)\theta_{KC} - \theta_{KU}. \end{aligned} \tag{9.25}$$

It is not a difficult job to show that Λ and Θ are both positive if the uncertainty sector is relatively more labor-intensive and negative if it is relatively more capital-intensive.[8]

Our next step is to simplify Eqs. (9.16), (9.18), (9.20), and (9.21) by eliminating every *aij* together with the λ and θ weighted sums of log a_{ij}'s. To this end, what we should note first is that U firms minimize unit cost $U_C = wa_{LC} + ra_{KC}$ for a given w and r. This obviously results in $dU_C = w\ da_{LU} + rda_{KU} = 0$, which can be rearranged in the following way:

$$(w\, a_{LU}/\mu_U)\,(da_{UC}/a_{LU}) + (r\, a_{KU}/\mu_U)\,(d\, a_{KU}/a_{KU}) = 0.$$

Obviously, this leads to the following nice equation of logarithmic differential.

$$\theta_{LU}\ d\ \log\ a_{UC} + \theta_{KU}\ d\ \log\ a_{KU} = 0. \tag{9.26}$$

Similarly, for C firms, we are able to derive the following nice equation.

$$\theta_{LC}\ d\ \log\ a_{LC} + \theta_{KU}\ d\ \log\ a_{KC} = 0. \tag{9.27}$$

Secondly, we may establish the relationship between factor price changes and factor proportions. To this end, we note the definition of the elasticity of substitution between factors in each sector. For U sector, we note the following definition:

[8] This relation among the signs of A, Λ, and Θ is now well established in the international trade literature. For example, see Jones (1965), Takayama (1972), and Ohyama (2011).

$$\sigma_U \equiv d \log (K_U/L_U)/d \log (w/r),$$

which can be rewritten as follows:

$$d \log K_U - d \log L_U = \sigma_U (d \log w - d \log r),$$

or equivalently,

$$(d \log K_U - d \log U) - (d \log L_U - d \log U)$$
$$= \sigma_U (d \log w - d \log r).$$

In terms of logarithmic differential, this implies the following:

$$\log a_{KU} - \log a_{LU} = \sigma_U (d \log w - d \log r). \quad (9.28)$$

For C sector, if the elasticity of substitution between factors is denoted by σ_C, then we can derive the following equation in a similar fashion.

$$\log a_{KC} - \log a_{LC} = \sigma_C (d \log w - d \log r). \quad (9.29)$$

If we combine those two equations (9.28) and (9.29) with Eqs. (9.26) and (9.27) and (9.22) and (9.23), we can obtain solutions for the $d \log a$'s of both sectors in the following way:

$$\log a_{LU} = -[\theta_{KU}/(1-\theta_R)]\sigma_U (d \log w - d \log r). \quad (9.30)$$

$$\log a_{KU} = [\theta_{LU}/(1-\theta_R)]\sigma_U (d \log w - d \log r). \quad (9.31)$$

$$\log a_{LC} = -\theta_{KC}\, \sigma_C (d \log w - d \log r). \quad (9.32)$$

$$\log a_{KC} = \theta_{LC}\, \sigma_C (d \log w - d \log r). \quad (9.33)$$

If we substitute these solutions into Eqs. (9.16), (9.17), (9.20), and (9.21), then we obtain the following set of equations:

$$\lambda_{LU}\, d \log U + \lambda_{LC}\, d \log C - \delta_L (d \log w - d \log r) = d \log L. \quad (9.34)$$

$$\lambda_{KU}\, d \log U + \lambda_{KC}\, d \log C - \delta_K (d \log w - d \log r) = d \log K. \quad (9.35)$$

$$\theta_{LU}\, d \log w + \theta_{KU}\, d \log r = d \log \mu_U - \theta_R (d \log \rho + d \log \gamma). \quad (9.36)$$

$$\theta_{LC}\, d \log w + \theta_{KC}\, d \log r = d \log p_C, \quad (9.37)$$

where δ_L and δ_K are the positive quantities, which are newly defined as follows.

$$\delta_L \equiv \lambda_{LU}\theta_{KU}\sigma_U/(1-\theta_R) + \lambda_{LC}\theta_{KC}\sigma_C.$$

$$\delta_K \equiv \lambda_{KU}\theta_{LU}\sigma_U/(1-\theta_R) + \lambda_{KC}\theta_{LC}\sigma_C.$$

Now, from Eq. (9.5) above, we find the following equation.

$$E[V_U{}'(\Pi_U)(\rho+\varepsilon)] = 0. \tag{9.38}$$

We note that total differentiation of this equation yields the following equation.

$$E[V_U{}''(\Pi_U)(\rho+\varepsilon)\, d\Pi_U] + E[V_U{}'(\Pi_U)]d\rho = 0. \tag{9.39}$$

Since $\Pi_U = (\rho + \varepsilon)\gamma U$ from Eq. (9.15), its total differentiation yields the following:

$$d\Pi_U = (\rho+\varepsilon)(\gamma d\,U + U\,d\gamma) + \gamma U\,d\rho.$$

By inserting this equation into (9.39) and rearranging it, we can obtain the following result:

$$-\alpha\, d\log U + \beta d\log\rho = \alpha d\log\gamma, \tag{9.40}$$

where α and β are the newly introduced quantities to be defined as follows:

$$\alpha \equiv -E[V_U{}''(\Pi_U)(\rho+\varepsilon)^2]\gamma U, \tag{9.41}$$

$$\beta \equiv -E[V_U{}''(\Pi_U)(\rho+\varepsilon)]\gamma U + E[V_U{}'(\Pi_U)]\rho. \tag{9.42}$$

Obviously, when U firms are risk-averse, α must be positive. The assumption that they exhibit decreasing absolute risk aversion is strong enough to determine the sign of β. To this end, let us define the absolute risk aversion function in the following way[9]:

$$R_U(\Pi_U) \equiv -V_U{}''(\Pi_C)/V_U{}'(\Pi_C). \tag{9.43}$$

Then, we can derive the following important result.

Lemma 9.2 (The Value of β) If $R_U{}'(\Pi_U) < 0$, then we find $\beta > 0$.

Proof First of all, let us rewrite (9.38) down below.

[9]The absolute risk aversion function of the form $R_U(\Pi_U) \equiv -V_U{}''(\Pi_C)/V_U{}'(\Pi_C)$ was first introduced by Pratt (1964) and Arrow (1970).

$$E[V_U'(\Pi_U)(\rho+\varepsilon)]=0.$$

Now, let us focus on a particular ε^* such that $V_U'(\Pi_U{}^*)(\rho+\varepsilon^*)=0$, where $\Pi_U{}^* = (\mu_U+\gamma\varepsilon^*)U - w\,a_{LU}U - r\,a_{KU}U$.

Since $\Pi_U - \Pi_U{}^* = \gamma(\varepsilon-\varepsilon^*)U$, we should have the following relation:

$$\Pi_U \gtreqless \Pi_U^* \Leftrightarrow \varepsilon \gtreqless \varepsilon^*(=-\rho)$$
$$\Leftrightarrow \rho+\varepsilon \gtreqless 0.$$

If $R_U'(\Pi_U)<0$, then we must have the following relation:

$$-V_U''(\Pi_C)/V_U'(\Pi_C) \equiv R_U'(\Pi_U) \lesseqgtr R_U'(\Pi_U^*)$$
$$\Leftrightarrow \Pi_U \gtreqless \Pi_U^*.$$

Accordingly, we should have the following result.

$$-[V_U''(\Pi_C)/V_U'(\Pi_C)](\rho+\varepsilon)$$
$$\leqq R_U'(\Pi_U{}^*)(\rho+\varepsilon) \text{ for any value of } \varepsilon,$$

which obviously implies the following:

$$V_U''(\Pi_C)(\rho+\varepsilon)$$
$$\geqq -R_U'(\Pi_U{}^*)V_U'(\Pi_C)(\rho+\varepsilon) \text{ for any value of } \varepsilon.$$

If we take expected values on both sides of this equation and note (9.28) above, we find

$$E[V_U''(\Pi_C)(\rho+\varepsilon)]$$
$$\geqq -R_U'(\Pi_U{}^*)E[V_U'(\Pi_C)(\rho+\varepsilon)]=0.$$

Clearly, this ensures that β is positive.

Q.E.D.

The system of equations of *relative or proportional change* for our uncertainty model is exhibited in (9.34)–(9.37) and (9.40). In matrix form, it can neatly be rewritten as follows.

$$\begin{bmatrix} \lambda_{LU} & \lambda_{LC} & 0 & -\delta_L & \delta_L \\ \lambda_{KU} & \lambda_{KC} & 0 & \delta_K & -\delta_K \\ -\alpha & 0 & \beta & 0 & 0 \\ 0 & 0 & \theta_R & \theta_{LU} & \theta_{KU} \\ 0 & 0 & 0 & \theta_{LC} & \theta_{KC} \end{bmatrix} \begin{bmatrix} d\log U \\ d\log C \\ d\log \rho \\ d\log w \\ d\log r \end{bmatrix} = \begin{bmatrix} d\log L \\ d\log K \\ \alpha d\log \gamma \\ d\log \mu_U - \theta_R d\log \gamma \\ d\log p_C \end{bmatrix}. \quad (9.44)$$

The subsystem consisting of the first two equations in the system (9.44) represents the commodity output factor endowment relationship and the subsystem consisting of the last two equations the factor price commodity price parameter relationship. The third equation, which links d log ρ to d log γ, tells us how the imposing presence of price risk contributes to make those two subsystems tightly connected and thus the dual feature shared by any two-by-two model more complicated than in the traditional, risk-free case.

In the special case in which U firms are just risk neutral, the quantities θ_R , α and β all vanish, so that (9.44) is simply reduced to the following:

$$\begin{bmatrix} \lambda_{LU} & \lambda_{LC} & -\delta_L & \delta_L \\ \lambda_{KU} & \lambda_{KC} & \delta_K & -\delta_K \\ 0 & 0 & \theta_{LU} & \theta_{KU} \\ 0 & 0 & 0 & \theta_{LC} \end{bmatrix} \begin{bmatrix} d \log U \\ d \log C \\ d \log w \\ d \log r \end{bmatrix} = \begin{bmatrix} d\log L \\ d\log K \\ d\log \mu_U \\ d\log p_C \end{bmatrix}. \quad (9.45)$$

As can easily be expected, the risk-neutral system (9.45) has essentially the same structure as the risk-free system with respect to the variables, parameters, and coefficients: the only difference is that the relative change in *expected* price of the uncertainty sector is now present as a variable.[10]

In this special system, the "financial" subsystem containing the last two equations may be entirely separated from the "physical" subsystem containing the first two equations in the following sense. The former subsystem itself is complete enough to determine the values $d\log w$ and $d\log r$ for given $d\log \mu_U$ and $d\log p_C$. As was seen above, however, such a simple separation of the total system into the two smaller subsystems is destined to vanish once "risk-averse players" come on the market stage. As everybody experiences in his daily life, risk aversion really matters and should not be neglected.

In order to understand the structure of the model more transparent, we believe that a more visual approach would be very instructive. Let us take a close look at Fig. 9.1. Then, on the left side, we can see a group of log differentials of *variables*—d log U, d log C, $d\log \rho$, $d\log w$, and $d\log r$. On the right side, we can see another group of log differentials of *parameters*—d log L, d log K, d log γ, d log μ_U, and d log p_C. The question of how and to what extent those two groups of log differentials are interlocked with each other is visually solvable in Fig. 9.1. It is also worthy of

[10]For the risk-free system of equations of relative change, see Jones (1965), for instance.

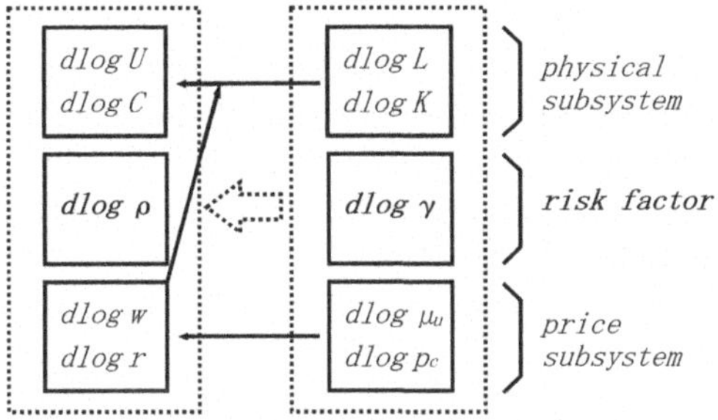

Fig. 9.1 The relationship between variables and parameters: risk neutrality versus risk aversion

reference to note that the model per se consists of the three subsystems—a *physical subsystem* related to the first two log differentials, a *price subsystem* related to the last two log differentials, and a still another *risk factor* placed between those two subsystems.

For convenience, the simple case of risk neutrality is illustrated by solid lines, and the more complicated case of risk aversion by dotted lines. As can easily be seen in Fig. 9.1, in the simple first case, the price subsystem is independent of the physical subsystem. Indeed, the values of $d \log w$ and $d \log r$ are dependent only on $d \log \mu_U$, and $d \log p_C$, thereby being determined exclusively within the price subsystem. In the complicated second case, however, a completely different situation would emerge. Then, as can be seen by a mixture of dotted lines, the aforementioned decomposition of the total system into the two subsystems would be no longer feasible, with the result that all the variables are dependent on all the parameters. No doubt, such decomposability would well explain the fundamental difficulty of comparative static analysis in the general case of risk aversion.

More specifically, let us take a careful look at Fig. 9.1 again. Then, three thick arrows taken together (➡) can show the strong relationship between variables and parameters in the specific case of risk neutrality. In a more general case of risk aversion, however, such clear relationship breaks down and should be changed to a more obscure one indicated by a single dotted line arrow. In short, the introduction of a risk factor may drastically change the structure of an international trade model.

9.4 The Relationship Between Factor Prices and Commodity Price Parameters: Reexamining the Validity of the Factor-Price Equalization Theorem

One of the most salient features of the non-stochastic two-sector, two-factor model is that under some reasonable conditions, factor prices are dependent only on commodity prices and on no other factors, which is particularly well known as the factor-price equalization theorem. We will show in this section, however, that such a nice relationship between factor prices and commodity prices cannot automatically be carried over to the world with price risk, except for the very special case in which firms in the uncertainty sector are just risk-neutral. In short, the presence of risk in the model is likely to throw the established order into disturbance.

In order to make the system (9.44) more manageable, let us attempt to reduce the number of variables from five to three. Making use of the last two equations in (9.44) (or equivalently, (9.36) and (9.37)), we may solve for $d \log w$ and $d \log r$ to derive the following equations.

$$\begin{aligned} &d \log w \\ &= (1/\Theta)[-\theta_{KC}\,\theta_R\, d \log \rho + \theta_{KC}\, d \log \mu_U \\ &\quad - \theta_{KC}\,\theta_R\, d \log \gamma - \theta_{KU}\, d \log p_C], \end{aligned} \tag{9.46}$$

$$\begin{aligned} &d \log r \\ &= (1/\Theta)[\theta_{LC}\,\theta_R\, d \log \rho - \theta_{LC}\, d \log \mu_U \\ &\quad + \theta_{LC}\,\theta_R\, d \log \gamma + \theta_{LU}\, d \log p_C]. \end{aligned} \tag{9.47}$$

In view of (9.22) and (9.23), we thereby obtain the following equation.

$$\begin{aligned} &d \log w - d \log r \\ &= (1/\Theta)[-\theta_R\, d \log \rho + d \log \mu_U \\ &\quad - \theta_R\, d \log \gamma - (1-\theta_R)\, d \log p_C]. \end{aligned} \tag{9.48}$$

As is quite clear from the above equations, the way in which factor prices are affected by changes in commodity price parameters should be dependent on in which direction and how much the risk aversion fee (ρ) responds to those changes. Note that $d \log \rho$ is linked to both $d \log L$ and $d \log K$ through both $d \log U$ and $d \log C$ (see the physical subsystem in (9.44)). This demonstrates the general dependence of factor price changes in the factor endowment. Therefore, the famous factor price equalization theorem no longer holds in our stochastic model, except for the special risk-neutral case in which θ_R, the risk factor's share, disappears from the stage.[11]

Substituting for $d \log w$ and $d \log r$ in (9.46) and (9.47) into the first three equations in the system (9.44) and rearranging them, we obtain the following matrix equation:

[11] Even in the risk-free case, factor prices may fail to be equalized when there are more factors than commodities. The merit and demerit of three-factor models in theory and history were systematically discussed by Jones (1971). In the light of such well-known result, the non-equalization of factor prices in the case of risk aversion could easily be understood if the risk-bearing fee (ρ) is interpreted as the payment for a *third factor to be counted.* Also see Samuelson (1949, 1953–1954).

$$\begin{bmatrix} \lambda_{LU} & \lambda_{LC} & \delta_L\theta_R/\Theta \\ \lambda_{KU} & \lambda_{KC} & -\delta_L\theta_R/\Theta \\ -\alpha & 0 & \beta \end{bmatrix} \begin{bmatrix} d\log U \\ d\log C \\ d\log \rho \end{bmatrix} =$$
$$\begin{bmatrix} d\log L + (\delta_L/\Theta)d\log \mu_U - (\delta_L\theta_R/\Theta)d\log\gamma - (\delta_L(1-\theta_R)/\Theta)d\log p_C \\ d\log K - (\delta_K/\Theta)d\log\mu_U + (\delta_K\theta_R/\Theta)d\log\gamma + (\delta_K(1-\theta_R)/\Theta)d\log p_C \\ \alpha d\log\gamma \end{bmatrix}. \tag{9.49}$$

Given changes in parameters, the values of d log U, d log C, and d log ρ are simultaneously determined in the reduced system (9.49). Then remarkably, the values of d log w and d log r are consequently determined by substituting for this d log ρ into (9.46) and (9.47). Let D be the determinant of the coefficient matrix in (9.49), namely,

$$D \equiv \begin{vmatrix} \lambda_{LU} & \lambda_{LC} & \delta_L\theta_R/\Theta \\ \lambda_{KU} & \lambda_{KC} & -\delta_L\theta_R/\Theta \\ -\alpha & 0 & \beta \end{vmatrix}.$$

Then, we find the value of D as follows.

$$D = \beta\Lambda + (\alpha\theta_R/\Theta)\,(\delta_L\,\lambda_{KC} + \delta_K\,\lambda_{LC}). \tag{9.50}$$

It is noted that D is positive if the uncertainty sector is relatively more labor-intensive, and negative if it is relatively more capital-intensive.

9.5 Changes in Factor Endowment: Reexamining the Validity of the Rybczynski Theorem

In this and following sections, we would like to investigate the comparative static properties of the uncertainty model described above more deeply by focusing on a change in one parameter only. First of all, we are interested in seeing what happens to the equilibrium variables of the system when the labor or capital endowment changes. Specifically, we wish to know whether or not the following theorem of Rybczynski remains unscathed by the introduction of price risk: an increase in the endowment of any specific factor leads to an expansion in the output of whichever sector is relatively more intensive in its use of that factor and to a contraction of the output of the other factor. For this point, see the fine piece of work by Rybczynski (1955).

We are anxious to know whether and to what extent the Rybczynski theorem remains valid by the presence of price risk (see Rybczynski 1955). In this connection, we will state and prove the following result.

Theorem 9.1 (The Rybczynski Theorem to be Reexamined)

1. Suppose $d \log L > 0$. Then, we have the following results.

$$d \log U > 0, d \log C \lesseqgtr 0, d \log \rho > 0, d \log w < 0, d \log r > 0$$

if the uncertainty sector is relatively more labor-intensive;

$$d \log U < 0, d \log C > 0, d \log \rho < 0, d \log w < 0, d \log r > 0$$

if it is relatively more capital-intensive.

2. Suppose $d \log L > 0$. Then we have the following results.

$$d \log U < 0, d \log C > 0, d \log \rho < 0, d \log w > 0, d \log r < 0$$

if the uncertainty sector is relatively more labor-intensive;

$$d \log U > 0, d \log C \lesseqgtr 0, d \log \rho > 0, d \log w > 0, d \log r < 0$$

if it is relatively more capital-intensive.

Proof Let all the parameters but $d \log L$ be zero in Eq. (9.49). Then we have the following matrix equation.

$$\begin{bmatrix} \lambda_{LU} & \lambda_{LC} & \delta_L \theta_R / \Theta \\ \lambda_{KU} & \lambda_{KC} & -\delta_L \theta_R / \Theta \\ -\alpha & 0 & \beta \end{bmatrix} \begin{bmatrix} \mathrm{dlog} U \\ \mathrm{dlog} C \\ \mathrm{dlog} \rho \end{bmatrix} = \begin{bmatrix} \mathrm{dlog} L \\ 0 \\ 0 \end{bmatrix}. \tag{9.51}$$

If we solve for the variables, it is not hard to derive the following equations.

$$d \log U = (\beta \lambda_{KC} / D)\, d \log L, \tag{9.52}$$

$$d \log C = -【(\beta \lambda_{KU}\, \Theta - \alpha \delta_K \theta_R) / D\Theta】\mathrm{d} \log \mathrm{L}, \tag{9.53}$$

$$d \log \rho = (\alpha \lambda_{KC} / D)\, d \log L. \tag{9.54}$$

By virtue of (9.46) and (9.47), we find the following equations.

$$d \log w = -(\theta_{KC}\, \theta_R / \Theta)\, d \log \rho, \tag{9.55}$$

$$d \log r = (\theta_{LC}\, \theta_R / \Theta)\, d \log \rho. \tag{9.56}$$

Keeping in mind that Θ and D are positive (or negative) if the uncertainty sector is relatively more labor (or capital)-intensive, Property (1) immediately follows from (9.52)–(9.56). Note that the sign of $d \log C$ is still indeterminate, depending the sign of the quantity $(\beta\lambda_{KU}\,\Theta - \alpha\delta_K\,\theta_R)$.

Now, focusing on $d \log K$, the proof of Property (2) proceeds in a similar way and is omitted. For a later discussion, we are only content to write the following formula:

$$d \log \rho = (\alpha\lambda_{LC}/D)\, d \log K. \tag{9.57}$$

This completes the proof.

Q.E.D.

Remarkably, Theorem 9.1 has several interesting implications in comparison with the non-stochastic results.

1. First of all, it is noted that the Rybczynski theorem may fail to hold for the certainty sector although it does hold for the uncertainty sector, provided that firms in the latter sector exhibit decreasing absolute risk aversion. More precisely, when the uncertainty sector is relatively more labor (or capital)-intensive, an increase in the labor endowment (or the capital endowment) does not necessarily result in a decrease in the output of the certainty sector. Surely, this looks a very delicate result. Since there is an intimate link between the Heckscher–Ohlin theorem and the Rybczynski theorem, this also implies the partial invalidity of the former theorem.[12]
2. Secondly, the risk-bearing fee varies in either direction in response to an increase in the factor endowment, depending on the factor intensity between the two sectors. This explains why the Rybczynski theorem may become partially invalid in the case of price risk. In fact, from the first two equations in (9.51), we find the following equations.

$$\lambda_{LU}\; d\; \log\; U + \lambda_{LC}\; d\; \log\; C = d\; \log\; L - (\delta_L\,\theta_R/\Theta)\; d\; \log\; \rho.$$

$$\lambda_{KU}\; d \log U + \lambda_{KC}\; d \log C = (\delta_K\,\theta_R/\Theta)\; d \log \rho.$$

Therefore, in case the uncertainty sector is relatively more labor intensive (i.e., Θ is positive), the increase in risk-bearing fee has a double effect in that it has an effect of decreasing L *and* another effect of increasing K. Taking those two effects into consideration, increasing ρ definitely shows the possibility that *both sectors simultaneously expand* as the result of the initial increase in L. In the light of (9.57), an analogous argument can effectively proceed for the implications of an increase in K.[13]

---

[12] This point was also noticed by Das (1977).

[13] It can be shown that $d \log L - (\delta_L\,\theta_R/\Theta)\, d \log \rho > 0$ if $d \log L > 0$ (see Eqs. 9.54 and 9.50). Therefore, the response of $d \log \rho$ to $d \log L$ makes a positive $d \log L$ smaller in the risk-free case, but not negative.

3. Thirdly, an increase in the endowment of any factor must cause the price of that factor to decline and the price of the other factor to rise, whatever the factor intensity between the two factors. This result can be well compared with the non-stochastic case in which factor prices are not affected at all by changes in factor endowments. In contrast to the risk-free world, the physical subsystem and the price subsystem are no longer separated in the risk-affected world: in plain English, the risk-bearing fee may serve as an "unlucky trouble maker" rather than a "lucky go-between."

So far, we have analyzed the compact of a change in only one parameter on equilibrium values. We now turn to a bit more complicated question, namely, the question of the impact of a *uniform relative change in the two parameters.*

Theorem 9.2 (The Impact of a Uniform Relative Change in Both L and K) Suppose $d\log L = d\log K > 0$. Then, we have the following results.

1. $d\ \log\ C > d\ \log\ U > 0.$
2. $d\ \log\ \rho > 0.$
3. $d\ \log\ w < 0$ and $d\ \log\ r > 0$

if the uncertainty sector is relatively more labor-intensive.

$$d\ \log\ w > 0 \text{ and } d\ \log\ r < 0$$

if it is relatively more capital-intensive.

Proof For convenience, let us put $d\log L = d\log K = d\log E$. Then, if we make use of (9.49) together with (9.46) and (9.47), it is a rather straightforward job to derive the following results.

$$d\ \log\ U = \frac{\beta\Lambda}{D}\ d\ \log\ E.$$

$$d\ \log\ C = \frac{\beta\Lambda\Theta + \alpha\theta_R(\delta_L + \delta_K)}{D\Theta}\ d\ \log\ E.$$

$$d\ \log\ U - d\ \log\ C = -\frac{\alpha\theta_R(\delta_L + \delta_K)}{D\Theta}\ d\ \log\ E.$$

$$d\ \log\ \rho = \frac{\alpha\Lambda}{D}\ d\ \log\ E.$$

$$d \log w = -\frac{\alpha\theta_{KL}\theta_R\Lambda}{D\Theta} d \log E.$$

$$d \log r = \frac{\alpha\theta_{KC}\theta_R\Lambda}{D\Theta} d \log E.$$

Using these formulas, the desired results follow immediately.
Q.E.D.

We say that a uniform relative change in the two factor endowments has a "neutral effect" if it does not alter the composition of outputs and the ratio of factor prices. Theorem 9.2 shows that as contrasted with the non-stochastic, constant-returns-scale economy, such neutral effect of factor endowments is no longer valid in our stochastic framework.

Let us suppose that both endowments expand at the same rate. Then, the uncertainty output grows at a lower rate than the certainty output. Besides, the price of whichever factor is used relatively more intensively in the uncertainty sector decreases and the price of the other sector increases. The reason why the uniform growth of both endowments produces such uneven effects on commodity outputs and factor prices is that it forces the risk-bearing fee to increase, which in turn affects the uncertainty sector relatively more unfavorably.[14]

In short, in the world of risk and uncertainty, people tend to exhibit risk aversion. Then, the burden of risk bearing is likely to play a trick on the economic stage.

9.6 Changes in (Expected) Commodity Price: the Stolper–Samuelson Theorem Reexamined

This section will consider the implications of a change in the expected price of the uncertainty commodity and of a change in the price of the certainty commodity. The famous theorem of Stolper and Samuelson, the one that is dual to the Rybczynski theorem, says that an increase in the price of any commodity results in an increase in the price of whichever factor is used relatively more intensively in the production of that commodity and in a decline in the price of the other factor. For this point, see the classical paper of Stolper and Samuelson (1941). Also see the related works of Samuelson (1949, 1953–1954, 1992), Rassekh and Thompson (1993), and Neary (2004).

Now, we will be especially interested in seeing how and to what extent the classical Stolper–Samuelson theorem can be carried over to the price risk world.

---

[14]As a corollary of Theorem 9.2, we see that there exists a sufficiently small positive value ξsuch that $d \log L = d \log K + \xi$, $d \log U < d \log C$ and the U sector is relatively more labor-intensive. Therefore, the magnification effect of factor supplies on commodity outputs cannot carry over to the case of risk aversion. It is recalled here that the magnification effect was first noticed by Jones (1965) for the risk-free case.jv

Theorem 9.3 (The Stopler–Samuelson Theorem to be Reexamined)

1. Suppose $d \log \mu_U > 0$. Then, we have the following results.

$$d\ \log\ U > 0, d\ \log\ C < 0, d\ \log\ \rho > 0, d\ \log\ w > 0, d\ \log\ r < 0$$

if the uncertainty sector is relatively more labor-intensive.

$$d\ \log\ U > 0, d\ \log\ C < 0, d\ \log\ \rho < 0, d\ \log\ w < 0, d\ \log\ r > 0$$

if it is relatively more capital-intensive.

2. Suppose $d \log p_C > 0$. Then we have the following results.

$$d\ \log\ U < 0, d\ \log\ C > 0, d\ \log\ \rho < 0, d\ \log\ w \lesseqgtr 0, d\ \log\ r > 0$$

if the uncertainty sector is relatively more labor-intensive.

$$d\ \log\ U < 0, d\ \log\ C > 0, d\ \log\ \rho < 0, d\ \log\ w > 0, d\ \log\ r \lesseqgtr 0$$

if it is relatively more capital-intensive.

Proof Taking advantage of 9.49, the proof of this theorem is almost parallel to that of Theorem 9.1 and is omitted here. We only record the following formulas.

1. For a change in *d log* μ_U, we note the following.

$$d \log \rho = \frac{\alpha(\delta_L \lambda_{KC} + \delta_K \lambda_{LC})}{D\Theta} d \log \mu_U, \tag{9.58}$$

$$d \log \mu_U - \theta_R \ d\log\rho = \frac{\beta\Lambda}{D} d \log \mu_U. \tag{9.59}$$

2. For a change in *d log* p_C, we note the following.

$$d \log \rho = -\frac{\alpha(1-\theta_R)(\delta_L \lambda_{KC} + \delta_K \lambda_{LC})}{D\Theta} d \log p_C, \tag{9.60}$$

$$d\ \log\ r - d\ \log p_C = \frac{\beta(1-\theta_R)\lambda_{LC}\Lambda}{D\Theta} d \log p_C, \tag{9.61}$$

$$d\ \log\ p_C - d\ \log\ w = \frac{\beta(1-\theta_R)\lambda_{KC}\Lambda}{D\Theta} d \log p_C, \tag{9.62}$$

Q.E.D.

In what follows, we can give some interesting interpretations to Theorem 9.2.

1. First of all, we see that although the Stolper–Samuelson Theorem is valid for the case with changes in μ_U, it may fail to be valid for the case with changes in p_C. In

fact, an increase in p_C does not always lead to a decline in w, namely, the price of the factor L that is less intensively used in C production. The key to such a non-symmetric result lies in the fact that the trouble-making ρ is present in the stochastic world, and that it rises when μ_U rises, but falls when p_C rises.

2. To see this point more sharply, on the one hand, we let all the parameters but $d \log \mu_U$ be zero in the last equation in (9.44) above and rearrange them slightly to obtain the following:

$$\theta_{LU} d \log w + \theta_{KU} d \log r = d \log \mu_U - \theta_R d \log \rho,$$
$$\theta_{LC} d \log w + \theta_{KC} d \log r = 0.$$

In the light of (9.59), the term ($d \log \mu_U - \theta_R\, d \log \rho$) is positive whenever $d \log \mu_U$ is so, whatever the intersecting factor intensity condition (note here that the quantities D, Λ, and Θ are either all positive or all negative). Consequently, the effect of the positive response of $d \log \rho$ to $d \log \mu_U$ is to make a positive $d \log \mu_U$ merely smaller, but not negative, with the result that the Stotper–Samuelson result continues to hold in this case.

3. On the other hand, we let all the parameters but $d \log p_C$ be zero in the last two equations in (9.44). Then, we find the following:

$$\theta_{LU} d \log w + \theta_{KU} d \log r = -\theta_R d \log \rho.$$
$$\theta_{LC} d \log w + \theta_{KC} d \log r = d \log p_C.$$

Therefore, a decreasing ρ accompanied with an increasing p_C has the same effect on w and r, as would an increasing μ_U. This tells us that there is a possibility that contrary to the Stolper–Samuelson theorem, both w and r go in the same direction as p_C.[15]

4. In short, the Stolper–Samuelson theorem is true, nice and strong in the risk-free world. When any kind of risk enters the world, however, we would have an entirely different picture. Then, the risk factor could possibly play as a "trouble maker." No matter how annoying it may be, it is the reality. Surely, we have to face up to it.

[15] Even so, w and r increase *at different rates*, depending on the factor intensity between the two sectors. As can readily be seen by eqs. (9.61) and (9.62) above, w rises at a faster rate than r (i.e., the wage-rental ratio rises) if the U sector is relatively more labor-intensive, and conversely, if it is relatively more capital-intensive.

Theorem 9.4 (The Impact of a Uniform Relative Change in Both μ_U and p_C) Suppose $d \log \mu_U = d \log p_C > 0$. Then, we have the following results.

1. $d \log U > 0, d \log C < 0$.
2. $d \log \rho > 0$.
3. $d \log w > 0, d \log r \gtreqless 0, d \log w > d \log r$

   if the uncertainty sector is relatively more labor-intensive.

   $$d \log w \gtreqless 0, d \log r > 0, d \log r > d \log w$$

   if it is relatively more capital-intensive.

Proof The proof of this theorem is quite analogous to that of Theorem 9.2 and is omitted.

Q.E.D.

We say that a uniform relative change in both μ_U and p_C has a "neutral effect" if it does not influence the output ration and the factor price ratio. As a dual of Theorem 9.2, Theorem 9.4 demonstrates such a neutral effect of (expected) commodity prices does not hold in the presence of price uncertainty, which is in marked contrast with the risk-free case. More specifically, we can derive following results.

1. Suppose that both μ_U and p_C rise at the same rate. Then, first of all, the uncertainty output U must increase, but the certainty output C must decrease.
2. Secondly, the price of whichever factor is more intensively used in the production of U output must rise at a faster rate than the price of the other factor, with the possibility that the latter price might even fall.
3. Note that risk-bearing fee is conspicuously present in the risk aversion case and positively responsive to the uniform commodity price change. This is the reason why the uniform change affects both physical and financial variables unevenly as stated above.[16]

[16] Theorem 9.4 clearly implies that the magnification effect a la Jones (1965) of commodity prices on factor prices cannot be extended to the world with price uncertainty. In a sense, the introduction of uncertainty into the model may play as a sort of "trouble maker."

9.7 Changes in Price Risk: Entering the New World of Uncertainty

We turn to the final important problem of this chapter. Now, we are concerned with the question of the implications of a change in price uncertainty for the equilibrium values of the model. We feel as if we were getting into the New World of Risk and Uncertainty. Hopefully, our approach will shed a new light on the old duality feature of the international economy.

Theorem 9.5 (Changes in Price Risk) Suppose $d \log \gamma > 0$. Then, we have the following results.

1. $d \log U < 0, d \log C > 0, d \log (\rho + \gamma) > 0, d \log w < 0, d \log r > 0$ if the uncertainty sector is relatively more labor-intensive.
2. $d \log U < 0, d \log C > 0, d \log (\rho + \gamma) > 0, d \log w > 0, d \log r < 0$ if it is relatively more labor-intensive.

Proof As was mentioned above, we start with the system (9.49). Now, limiting our attention to $d \log \gamma$, the proof is rather straightforward. We only write the following formulas.

$$d \log \rho = -\frac{\alpha\{\theta_R(\delta_L\lambda_{KC} + \delta_K\lambda_{LC}) - \Lambda\Theta\}}{D\Theta} d \log \gamma. \tag{9.63}$$

$$d \log \rho + d \log \gamma = \frac{(\alpha + \beta)\Lambda}{D} d \log \gamma. \tag{9.64}$$

$$d \log w = -\frac{\theta_{KC}\theta_R}{\Theta} (d \log \rho + d \log \gamma). \tag{9.65}$$

$$d \log r = \frac{\theta_{LC}\theta_R}{\Theta} (d \log \rho + d \log \gamma). \tag{9.66}$$

Q.E.D.

We note that Theorem 9.5 is quite interesting in several aspects.

1. U sector may be relatively more labor-intensive or more capital-intensive) than C sector. In spite of the difference of the factor intensity condition between the two sectors, an increase in price risk must lead to a contraction of U sector and to an expansion of C sector. This agrees with common sense.
2. Note that when the price risk measured by γ goes up, the bearing fee ρ may go in either direction. However, as can clearly be understood, the sign of the "combined risk burden," which is represented by the sum of the "subjective risk burden" ($d \log \rho$) and the "objective risk burden" ($d \log \gamma$), must definitely be positive.
3. The abovementioned result plays a crucial role in determining the signs of $d \log w$ and $d \log r$, as is seen in (9.65) and (9.66). In fact, the price risk increase results

in a decline in the price of whichever factor is relatively more intensively used, and as a linking reaction, a rise in the price of the other factor,

4. Now, comparison of Theorems 9.3 and 9.5 shows that as may naturally be expected, the implications of a change in μ_U and those of a change in γ are antithetical in the case of risk aversion.

As the saying goes, a wise man keeps clear of danger. The reality, however, is different from the saying, and many men are not so wise. We have to exert all our energy for filling in such a gap between the ideal and the reality.

9.8 A Unification of Three Streams of Economic Theories: General Equilibrium, International Trade, and Risk and Uncertainty

This chapter, from the beginning to the end, has been an ambitious attempt to unify the three main three streams of economic theories into one.

The first stream of economic theory is *general equilibrium theory,* which has been developed a great deal by a group of shining superstars including Lionel W. McKenzie, Kenneth J. Arrow, and Gerald Debreu. By combining the initials of those three surnames, we may be allowed to call this theory "the M-A-D theory." We are not quite sure of "how mad" it really is. At least, we could safely say that it is one of the most beautiful theories economists have ever produced over hundred years.[17]

The second stream is *theory of international trade*, which has greatly prompted by a group of smart brains including Ronald W. Jones, Murray C. Kemp, and J. Bhagwati. By combining the initials of those three surnames, we may be allowed to say this theory "the J-K-B theory." We are not quite certain of how "smart" it truly is. At least, we could say that it is one of the smartest theories international economists ever generated for so many years. For this point, see Flemming et al. (1977), Bhagwati et al. (1971), Deardorff (1984), Flam and Flanders (1991), and Feenstra (2004).[18]

The third stream is *theory of risk and uncertainty*, which has a very long history but, to our regret, has been treated rather lightly until recently. It is in the 1970s and

[17]Personally speaking, Yasuhiro Sakai was once McKenzie's student at the University of Rochester. McKenzie's teaching style was very unique in that he sometimes pondered for a while with white chalk on his lips. His triumphant face after struggling to prove the existence theorem of general equilibrium by means of the powerful Kakutani fixed point theorem was an unforgettable episode in Sakai's life. So, McKenzie was rightly nicknamed "Professor Fixed Point" by many students.

[18]Dr. Makoto Tawada, who received his Ph.D. degree at Australian National University, is one of Sakai's best friends in Regional Economic Association. Tawada used to tell Sakai that Jones, Kemp, and Bhagwati jointly deserved Nobel Prize in Economic Science. Besides, Kemp himself once told Sakai that McKenzie should be awarded the Nobel Economics Prize.

afterward that a modern approach to risk and uncertainty came out into the economics profession, with superstars being Kenneth J Arrow, George A. Akerlof, Michael Spence, and George J. Stiglitz. Since the initials of those four superstars are either "A" or "S," we may be allowed to call this theory "the AA-SS theory." We are not so confident of how "great" it certainly is. We do believe, however, that it constitutes one of the most outstanding theories over so many decades or even so many centuries.[19]

To tell the truth, it is not an easy job to combine those three streams of economic theories into one. As the saying goes, however, life is a challenge, and an adventure as well. We believe that although this is the right way to go, it could be only the first step to go. Perhaps, there would be a variety of ways to proceed. There remain so many unsolved problems, requiring future research. In the world of uncertainty, there should be one thing certain: that is, where there is a will, there is a way. We sincerely hope that our work in this book should be continued to the next generations, thus shedding a new light for further development of the theories of games, decisions, and markets.

References

Arrow KJ (1970) Essays in the theory of risk-bearing. North Holland, Amsterdam

Batra RN (1974) Resource allocation in a general equilibrium model of production under uncertainty. J Econ Theory 8:59–63

Batra RN (1975a) Production uncertainty and the Heckscher-Ohlin theorem. Rev Econ Stud 42: 259–268

Batra RN (1975b) The pure theory of international trade under uncertainty. Macmillan, London

Bhagwati J (1964) The pure theory of international trade: a survey. Econ J 74:1–84

Bhagwati JN, Jones RW, Mundell RA, Vanek J (eds) (1971) Trade, balance of payments, and growth: papers in international economics in honor of Charles P. Kindleberger. North Holland, Amsterdam

Caves RE, Jones RW (1973) World trade and payments. Little Brown, Boston

Coppock JD (1962) International economic instability: the experience after Word War II. McGraw Hill, New York

Das SK (1977) Uncertainty and the Heckscher-Ohlin theorem: a comment. Rev Econ Stud 44:189–190

Deardorff A (1984) Testing trade theories and predicting trade flows. In: Kenen PB (ed) The handbook of international economics, vol 1. North Holland, Amsterdam

Feenstra RC (2004) The Heckscher-Olin model. In: Jones RW (ed) Advanced international trade: theory and evidence. Princeton University Press, Princeton

Flam H, Flanders MJ (1991) Heckcher-Ohlin trade theory. MIT Press, Cambridge

Flemming JS, Turnovsky SJ, Kemp MC (1977) On the theory of numeraire and certainty price in general equilibrium models. Rev Econ Stud 44:573–583

[19] It was Oskar Morgenstern who recommended me at the University of Pittsburgh to do research in the theory of risk and uncertainty. By chance, Sakai's initial is no less than "S," one of the lucky letters for risk researchers. Thus, I have been greatly indebted to Morgenstern for very kind suggestions.

Jones RW (1965) The structure of simple general equilibrium models. J Polit Econ 53:557–572
Jones RW (1971) A three-sector model in theory, trade, and history. In: Bhagwati, Jones, Vanek (eds) Trade, balance of payments, and growth. North-Holland, Amsterdam-London
Kemp MC (1969) The pure theory of international trade and investment. Prentice Hall, New Jersey
Kemp MC (1976) Three topics in the theory of international trade. North Holland, Amsterdam
Komiya R, Amano A (1972) International economics (in Japanese). Iwanami Publishers, Tokyo.
Krugman P, Obstfeld M (2007) Resources and trade: the Heckscher-Ohlin model. In: Krugman P (ed) International economics: theory and trade. Addison Wesley, Boston, pp 67–92
Leamer EE (1995) The Heckscher-Ohlin model in theory and practice. Princeton studies in international finance 77, Princeton University Press. https://archive.org/details/heckscherohlinmo0000learn
Mayer W (1976) The Rybczynski, Stolper-Samuelson, and factor-price equalization theorems under price uncertainty. Am Econ Rev 66:797–808
Neary JP (2004) The Stolper-Samuelson theorem. Discussion paper, Centre for Economic Policy Research, London
Ohyama M (2011) International economics (in Japanese). Baihukan Publishers, Tokyo
Pratt JW (1964) Risk aversion in the small and in the large. Econometrica 32:122–136
Rassekh F, Thompson H (1993) Factor price equalization: theory and evidence. J Econ Integr Spring:1–32
Rybczynski TN (1955) Factor endowments and relative commodity prices. Economica 33:336–341
Sakai Y (1977) The theory of the firm under price uncertainty. Econ Stud Q 28:31–39
Sakai Y (1978) A simple general equilibrium model of production: comparative statics with price uncertainty. J Econ Theory 19(2):287–306
Sakai Y (1982) The economics of uncertainty (in Japanese). Yuhikaku publishers, Tokyo
Samuelson PA (1949) International factor price equalization once again. Economic J 59:181–197
Samuelson PA (1953–1954) Prices of factors and goods in general equilibrium. Rev Econ Stud 21: 1–20
Samuelson PA (1992) Factor price equalization by trade in joint and non-joint production. Rev Int Econ 1:1–9
Sandmo A (1971) On the theory of the competitive firm under price uncertainty. Am Econ Rev 65: 65–73
Stolper WE, Samuelson PA (1941) Protection and real wages. Rev Econ Stud 9:598–573
Takayama A (1972) International economics. Holt, Rinehart & Winston, New York

SPRINGER NATURE

GPSR Compliance

The European Union's (EU) General Product Safety Regulation (GPSR) is a set of rules that requires consumer products to be safe and our obligations to ensure this.

If you have any concerns about our products, you can contact us on ProductSafety@springernature.com

In case Publisher is established outside the EU, the EU authorized representative is:

Springer Nature Customer Service Center GmbH
Europaplatz 3
69115 Heidelberg, Germany

The manufacturer's authorised representative in the EU is Springer Nature Customer Service Centre GmbH, Europaplatz 3, 69115 Heidelberg, Germany. If you have any concerns regarding our products, please contact ProductSafety@springernature.com

Printed and bound by CPI Group (UK) Ltd, Croydon, CR0 4YY
01/10/2026
02225897-0006